Memoirs
of the
American Mathematical Society

Volume 233 • Number 1098 (fourth of 6 numbers) • January 2015

A Geometric Theory for Hypergraph Matching

Peter Keevash
Richard Mycroft

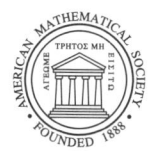

ISSN 0065-9266 (print)   ISSN 1947-6221 (online)

American Mathematical Society
Providence, Rhode Island

Library of Congress Cataloging-in-Publication Data

Keevash, Peter, 1978- author.
   A geometric theory for hypergraph matching / Peter Keevash, Richard Mycroft.
   pages cm. – (Memoirs of the American Mathematical Society, ISSN 0065-9266 ; number 1098)
   "January 2015, volume 233, number 1098 (fourth of 6 numbers)."
   Includes bibliographical references.
   ISBN 978-1-4704-0965-4 (alk. paper)
   1. Hypergraphs. 2. Matching theory. I. Mycroft, Richard, 1985- author. II. Title.
QA166.23.K44 2014
511'.5–dc23
                                                                                                                        2014033269
DOI: http://dx.doi.org/10.1090/memo/1098

Memoirs of the American Mathematical Society

This journal is devoted entirely to research in pure and applied mathematics.

Subscription information. Beginning with the January 2010 issue, *Memoirs* is accessible from www.ams.org/journals. The 2015 subscription begins with volume 233 and consists of six mailings, each containing one or more numbers. Subscription prices for 2015 are as follows: for paper delivery, US$860 list, US$688.00 institutional member; for electronic delivery, US$757 list, US$605.60 institutional member. Upon request, subscribers to paper delivery of this journal are also entitled to receive electronic delivery. If ordering the paper version, add US$10 for delivery within the United States; US$69 for outside the United States. Subscription renewals are subject to late fees. See www.ams.org/help-faq for more journal subscription information. Each number may be ordered separately; *please specify number* when ordering an individual number.

Back number information. For back issues see www.ams.org/bookstore.

Subscriptions and orders should be addressed to the American Mathematical Society, P. O. Box 845904, Boston, MA 02284-5904 USA. *All orders must be accompanied by payment.* Other correspondence should be addressed to 201 Charles Street, Providence, RI 02904-2294 USA.

Copying and reprinting. Individual readers of this publication, and nonprofit libraries acting for them, are permitted to make fair use of the material, such as to copy select pages for use in teaching or research. Permission is granted to quote brief passages from this publication in reviews, provided the customary acknowledgment of the source is given.

Republication, systematic copying, or multiple reproduction of any material in this publication is permitted only under license from the American Mathematical Society. Permissions to reuse portions of AMS publication content are handled by Copyright Clearance Center's RightsLink® service. For more information, please visit: http://www.ams.org/rightslink.

Send requests for translation rights and licensed reprints to reprint-permission@ams.org.

Excluded from these provisions is material for which the author holds copyright. In such cases, requests for permission to reuse or reprint material should be addressed directly to the author(s). Copyright ownership is indicated on the copyright page, or on the lower right-hand corner of the first page of each article within proceedings volumes.

Memoirs of the American Mathematical Society (ISSN 0065-9266 (print); 1947-6221 (online)) is published bimonthly (each volume consisting usually of more than one number) by the American Mathematical Society at 201 Charles Street, Providence, RI 02904-2294 USA. Periodicals postage paid at Providence, RI. Postmaster: Send address changes to Memoirs, American Mathematical Society, 201 Charles Street, Providence, RI 02904-2294 USA.

© 2014 by the American Mathematical Society. All rights reserved.
Copyright of individual articles may revert to the public domain 28 years
after publication. Contact the AMS for copyright status of individual articles.
This publication is indexed in *Mathematical Reviews*®, *Zentralblatt MATH*, *Science Citation Index*®, *Science Citation Index*^{TM}*-Expanded*, *ISI Alerting Services*^{SM}, *SciSearch*®, *Research Alert*®, *CompuMath Citation Index*®, *Current Contents*®/*Physical, Chemical & Earth Sciences*. This publication is archived in *Portico* and *CLOCKSS*.
Printed in the United States of America.

∞ The paper used in this book is acid-free and falls within the guidelines
established to ensure permanence and durability.
Visit the AMS home page at http://www.ams.org/

10 9 8 7 6 5 4 3 2 1   20 19 18 17 16 15

Contents

Chapter 1. Introduction — 1
 1.1. Space barriers and divisibility barriers — 1
 1.2. Tetrahedron packings — 3
 1.3. A multipartite Hajnal-Szemerédi theorem — 4
 1.4. Algorithmic aspects of hypergraph matchings — 5
 1.5. Notation — 6

Chapter 2. Results and examples — 7
 2.1. Almost perfect matchings — 7
 2.2. Partite systems — 9
 2.3. Lattice-based constructions — 10
 2.4. Perfect matchings — 12
 2.5. Further results — 15
 2.6. Outline of the proofs — 16

Chapter 3. Geometric Motifs — 19

Chapter 4. Transferrals — 23
 4.1. Irreducibility — 23
 4.2. Transferral digraphs — 24
 4.3. Completion of the transferral digraph — 28

Chapter 5. Transferrals via the minimum degree sequence — 31

Chapter 6. Hypergraph Regularity Theory — 39
 6.1. Hypergraph regularity — 39
 6.2. The Regular Approximation Lemma — 40
 6.3. The hypergraph blowup lemma — 41
 6.4. Reduced k-systems — 42
 6.5. Proof of Lemma 5.5 — 45

Chapter 7. Matchings in k-systems — 51
 7.1. Fractional perfect matchings — 51
 7.2. Almost perfect matchings — 57
 7.3. Perfect matchings — 61

Chapter 8. Packing Tetrahedra — 69
 8.1. Packing to within a constant — 71
 8.2. Properties of index vectors — 72
 8.3. Divisibility barriers with two parts — 73
 8.4. Divisibility barriers with more parts — 78

8.5.	The main case of Theorem 1.1	81
8.6.	The case when 8 divides n	84
8.7.	Strong stability for perfect matchings	85

Chapter 9. The general theory   89
    Acknowledgements   91

Bibliography   93

Abstract

We develop a theory for the existence of perfect matchings in hypergraphs under quite general conditions. Informally speaking, the obstructions to perfect matchings are geometric, and are of two distinct types: 'space barriers' from convex geometry, and 'divisibility barriers' from arithmetic lattice-based constructions. To formulate precise results, we introduce the setting of simplicial complexes with minimum degree sequences, which is a generalisation of the usual minimum degree condition. We determine the essentially best possible minimum degree sequence for finding an almost perfect matching. Furthermore, our main result establishes the stability property: under the same degree assumption, if there is no perfect matching then there must be a space or divisibility barrier. This allows the use of the stability method in proving exact results. Besides recovering previous results, we apply our theory to the solution of two open problems on hypergraph packings: the minimum degree threshold for packing tetrahedra in 3-graphs, and Fischer's conjecture on a multipartite form of the Hajnal-Szemerédi Theorem. Here we prove the exact result for tetrahedra and the asymptotic result for Fischer's conjecture; since the exact result for the latter is technical we defer it to a subsequent paper.

Received by the editor September 19, 2011, and, in revised form, March 8, 2013.
Article electronically published on May 19, 2014.
DOI: http://dx.doi.org/10.1090/memo/1098
2010 *Mathematics Subject Classification.* Primary 05C65, 05C70.
Key words and phrases. hypergraphs, perfect matchings.
Research supported in part by ERC grant 239696 and EPSRC grant EP/G056730/1.

©2014 American Mathematical Society

CHAPTER 1

Introduction

Hypergraph matchings[1] provide a general framework for many important Combinatorial problems. Two classical open problems of this nature are the question of whether there exist designs of arbitrary strength, and Ryser's conjecture that every Latin square of odd order has a transversal; these are both equivalent to showing that some particular hypergraphs have perfect matchings. Furthermore, matchings are also an important tool for many practical questions, such as the 'Santa Claus' allocation problem (see [2]). However, while Edmonds' algorithm [12] provides an efficient means to determine whether a graph has a perfect matching, the decision problem is NP-complete in k-graphs for $k \geq 3$ (it is one of Karp's original 21 NP-complete problems [21]). Thus we do not expect a nice characterisation, so we concentrate on natural sufficient conditions for finding a perfect matching.

One natural hypergraph parameter that is widely considered in the literature is the *minimum degree* $\delta(G)$, which is the largest number m such that every set of $k-1$ vertices is contained in at least m edges of G. What is the minimum degree threshold for finding a perfect matching? (We assume that $k|n$, where $n = |V|$.) For graphs ($k = 2$) a straightforward greedy argument shows that the threshold is $n/2$ (or one can deduce it from Dirac's theorem [11], which states that the same threshold even gives a Hamilton cycle). The general case was a long-standing open problem, finally resolved by Rödl, Ruciński and Szemerédi [49], who determined the threshold precisely for large n: it is $n/2 - k + C$, where $C \in \{1.5, 2, 2.5, 3\}$ depends on arithmetic properties of k and n. There is a large literature on minimum degree problems for hypergraphs, see e.g. [1, 6, 7, 10, 11, 14, 19, 25–28, 30, 31, 33, 35, 36, 38–40, 44, 47–49, 57] and the survey by Rödl and Ruciński [46] for details.

1.1. Space barriers and divisibility barriers

To motivate the results of this paper it is instructive to consider the extremal examples for the minimum degree problems. Consider a graph G_1 on n vertices whose edges are all pairs incident to some set S of size $n/2-1$. Then $\delta(G_1) = n/2-1$, and G_1 has no perfect matching, as each edge of a matching M uses a vertex in S, so $|M| \leq |S|$; we say that G_1 has a *space barrier*. Now suppose $n/2$ is odd and consider a graph G_2 on n vertices consisting of two disjoint complete graphs of size $n/2$. Then $\delta(G_2) = n/2 - 1$, and G_2 has no perfect matching, but for the different reason that edges have even size; we say that G_2 has a *divisibility barrier*.

While these two examples are equally good for graphs, for general k-graphs a separation occurs. For G_1 we take a k-graph whose edges are all k-tuples incident to some set S of size $n/k - 1$; this satisfies $\delta(G_1) = n/k - 1$. For G_2 we take a k-graph

---

[1] A *hypergraph* G consists of a vertex set V and an edge set E, where each edge $e \in E$ is a subset of V. We say G is a *k-graph* if every edge has size k. A *matching* M is a set of vertex disjoint edges in G. We call M *perfect* if it covers all of V.

whose edges are all k-tuples that have an even size intersection with some set S such that $n/2 - 1 \le |S| \le (n+1)/2$ and $|S|$ is odd; this satisfies $\delta(G_1) = n/2 - k + C$, being one of the extremal constructions in the result of [**49**] mentioned above. One should also note that space is a robust obstruction to matchings, in that the size of a maximum matching in G_1 may be decreased by decreasing $|S|$, whereas divisibility is not robust, in the sense that any construction similar to G_2 has an almost perfect matching. In fact, Rödl, Ruciński and Szemerédi [**48**] showed that the minimum degree threshold for a matching of size $n/k - t$ with $t \ge k - 2$ is $n/k - t$. Thus we see a sharp contrast between the thresholds for perfect matching and almost perfect matching.

The main message of this paper is that space and divisibility are the determining factors for perfect matchings in hypergraphs under quite general conditions, and that these factors are inherently geometric. The first part of the geometric theory was anticipated by a result on fractional perfect matchings in [**48**] that we generalise here. The key point is that fractional perfect matchings correspond to representing a constant vector as a convex combination of edge vectors, so non-existence of fractional perfect matchings can be exploited in geometric form via separating hyperplanes. Furthermore, the fractional problem has bearing on the original problem through standard 'regularity machinery', which converts a fractional solution into a matching covering all but $o(n)$ vertices for large n.

The second part of the theory is to understand when the number of uncovered vertices can be reduced to a constant independent of n. The idea here can only be properly explained once we have described the regularity embedding strategy, but a brief summary is as follows. Firstly, the $o(n)$ uncovered vertices arise from imbalances created when converting from the fractional to the exact solution. Secondly, the possible 'transferrals' of imbalances can be expressed geometrically by defining an appropriate polyhedron and testing for a ball around the origin of some small constant radius; this can also be understood in terms of separating hyperplanes. Thus the first two parts of the theory are problems of convex geometry, which correspond to space barriers.

The third part of the theory concerns divisibility barriers, which determine when the number of uncovered vertices can be reduced from a constant to zero. We will see that in the absence of space barriers, perfect matchings exist except in hypergraphs that are structurally close to one of a certain class of arithmetic constructions defined in terms of lattices in $\mathbb{Z}^d$ for some $d \le k$. Furthermore, since the constructions with space or divisibility barriers do not have perfect matchings, in a vague sense we have 'the correct theory', although this is not a precise statement because of additional assumptions. Our theory is underpinned by the 'strong' hypergraph regularity theory independently developed by Gowers [**17**] and Rödl et al. [**15, 43, 50, 52**], and the recent hypergraph blowup lemma of Keevash [**22**]. Fortunately, this part of the argument is mostly covered by existing machinery, so the majority of this paper is devoted to the geometric theory outlined above.

To formulate precise results, we introduce the setting of simplicial complexes with minimum degree sequences, which is a generalisation of the minimum degree condition previously considered. In this setting our main theorems (stated in Section 2.4) give minimum degree sequences that guarantee a perfect matching for hypergraphs that are not close to a lattice construction. These minimum degree sequences are best possible, and furthermore have the 'stability' property that, unless the hypergraph is structurally close to a space barrier construction, one can find a perfect matching even with a slightly smaller degree sequence. We defer the statements until we have given the necessary definitions in the next chapter. For now we want to emphasise the power of this framework by describing its application to the solutions of two open problems on packings. Suppose G is a k-graph on n vertices and H is a k-graph on h vertices, where $h|n$ (we think of h as fixed and n as large). An H-*packing* in G is a set of vertex-disjoint copies of H inside G; it is *perfect* if there are n/h such copies, so that every vertex of G is covered. In the case when H is a single edge we recover the notion of (perfect) matchings.

As for matchings we have the natural question: what is the minimum degree threshold for finding a perfect H-packing? Even for graphs, this is a difficult question with a long history. One famous result is the Hajnal-Szemerédi theorem [18], which determines the threshold for the complete graph K_r on r vertices: if $r|n$ and $\delta(G) \geq (r-1)n/r$ then G has a K_r-packing, and this bound is best possible. It is interesting to note that this is essentially the same threshold as for the Turán problem [58] for K_{r+1}, i.e. finding a single copy of the complete graph with one more vertex. The perfect packing problem for general graphs H was essentially solved by Kühn and Osthus [31], who determined the threshold for large n up to an additive constant $C(H)$. The precise statement would take us too far afield here, but we remark that the threshold is determined by either space or divisibility barriers, and that the dependence on the chromatic number of H continues a partial analogy with Turán problems. We refer the reader to their survey [32] for further results of this type.

1.2. Tetrahedron packings

For hypergraph packing problems, the natural starting point is the tetrahedron K_4^3, i.e. the complete 3-graph on 4 vertices. Here even the asymptotic existence threshold is a long-standing open problem; this is an important test case for general hypergraph Turán problems, which is a difficult area with very few known results (see the survey by Keevash [22]). In light of this, it is perhaps surprising that we are able here to determine the tetrahedron packing threshold for large n, not only asymptotically but precisely. One should note that the two problems are not unrelated; indeed Turán-type problems for the tetrahedron are required when showing that there are no divisibility barriers (but fortunately they are more tractable than the original problem!) The extremal example for the perfect tetrahedron packing problem is by no means obvious, and it was several years after the problem was posed by Abbasi (reported by Czygrinow and Nagle [8]) that Pikhurko [44] provided the optimal construction (we describe it in Chapter 8). Until recently, the best published upper bounds, also due to Pikhurko [44], were $0.8603\ldots n$ for the perfect packing threshold and $3n/4$ for the almost perfect packing threshold. More recent upper bounds for the perfect packing threshold are $4n/5$ by Keevash and Zhao (unpublished) and $(3/4+o(1))n$ independently by Keevash and Mycroft (earlier manuscripts of this paper) and by Lo and Markström [35] (posted online very recently). It is instructive to contrast the 'absorbing technique' used in the proof of [35] with the approach here; we will make some remarks later to indicate why our methods seem more general and are able to give the exact result, which is as follows.

THEOREM 1.1. *There exists n_0 such that if G is a 3-graph on $n \geq n_0$ vertices such that $4 \mid n$ and*

$$\delta(G) \geq \begin{cases} 3n/4 - 2 & \text{if } 8 \mid n \\ 3n/4 - 1 & \text{otherwise,} \end{cases}$$

then G contains a perfect K_4^3-packing. This minimum degree bound is best possible.

The minimum degree bound of Theorem 1.1 is best possible. Indeed, consider a 3-graph G whose vertex set V is the disjoint union of sets A, B, C and D whose sizes are as equal as possible with $|A|$ odd. The edges of G are all 3 tuples except those

(i) with all vertices in A,
(ii) with one vertex in A and the remaining two vertices in the same vertex class, or
(iii) with one vertex in each of B, C and D.

There is then no perfect K_4^3-packing in G (see Proposition 8.1 for details), but $\delta(G)$ is equal to $3n/4 - 3$ if $8 \mid n$, and $3n/4 - 2$ otherwise.

1.3. A multipartite Hajnal-Szemerédi theorem

Our second application is to a conjecture of Fischer [14] on a multipartite form of the Hajnal-Szemerédi Theorem. Suppose $V_1, \ldots, V_k$ are disjoint sets of n vertices each, and G is a k-partite graph on vertex classes $V_1, \ldots, V_k$ (that is, G is a graph on $V_1 \cup \cdots \cup V_k$ such that no edge of G has both vertices in the same V_j). Then we define the *partite minimum degree* of G, denoted $\delta^*(G)$, to be the largest m such that every vertex has at least m neighbours in each part other than its own, i.e.

$$\delta^*(G) = \min_{i \in [k]} \min_{v \in V_i} \min_{j \in [k] \setminus \{i\}} |N(v) \cap V_j|,$$

where $N(v)$ denotes the neighbourhood of v. Fischer conjectured that if $\delta^*(G) \geq (k-1)n/k + 1$ then G has a perfect K_k-packing (actually his original conjecture did not include the $+1$, but this stronger conjecture is known to be false). The case $k = 2$ of the conjecture is an immediate corollary of Hall's Theorem, whilst the cases $k = 3$ and $k = 4$ were proved by Magyar and Martin [38] and Martin and Szemerédi [40] respectively. Also, Csaba and Mydlarz [6] proved a weaker version of the conjecture in which the minimum degree condition has an error term depending on k. The following theorem, an almost immediate corollary of our results on hypergraph matchings, gives an asymptotic version of this conjecture for any k.

THEOREM 1.2. *For any k and $c > 0$ there exists n_0 such that if G is a k-partite graph with parts $V_1, \ldots, V_k$ of size $n \geq n_0$ and $\delta^*(G) \geq (k-1)n/k + cn$, then G contains a perfect K_k-packing.*

This asymptotic result was also proven independently and simultaneously by Lo and Markström [36] in another application of the 'absorbing technique'. As with Theorem 1.1, by considering the near-extremal cases of the conjecture using the 'stability property' of our main theorem, we are able to prove an exact result in [26], namely that Fischer's conjecture holds for any sufficiently large n. However, this stability analysis is lengthy and technical, so we prefer to divest this application from the theory developed in this paper.

As mentioned above, we will deduce both Theorems 1.1 and 1.2 from a general framework of matchings in simplicial complexes. These will be formally defined in the next chapter, but we briefly indicate the connection here. For Theorem 1.1 we consider the 'clique 4-complex', with the tetrahedra in G as 4-sets, G as 3-sets, and all smaller sets; for Theorem 1.2 we consider the 'clique k-complex', with the j-cliques of G as j-sets for $j \leq k$. In both cases, the required perfect packing is equivalent to a perfect matching using the highest level sets of the clique complex.

1.4. Algorithmic aspects of hypergraph matchings

As described earlier, the decision problem of whether a k-graph H has a perfect matching is NP-complete for $k \geq 3$, motivating our consideration of the minimum degree which guarantees that H contains a perfect matching. Another natural question to ask is for the minimum-degree condition which renders the decision problem tractable. That is, let $PM(k, \delta)$ denote the problem of deciding whether a k-graph H on n vertices (where $k \mid n$) with $\delta(H) \geq \delta n$ contains a perfect matching. The result of Karp [21] mentioned earlier shows that $PM(k, 0)$ is NP-complete. On the other hand, the theorem of Rödl, Ruciński and Szemerédi [49] described earlier shows that any sufficiently large k-graph on n vertices with $\delta(H) \geq n/2$ contains a perfect matching, so $PM(k, \delta)$ can be solved in constant time for $\delta \geq 1/2$ by an algorithm which simply says 'yes' if n is sufficiently large, and checks all possible matchings by brute force otherwise.

This question was further studied by Szymańska [53], who proved that for $\delta < 1/k$ the problem $PM(k, 0)$ admits a polynomial-time reduction to $PM(k, \delta)$, and so $PM(k, \delta)$ is NP-complete for such δ. In the other direction, Karpiński, Ruciński and Szymańska [34] proved that there exists a constant $\varepsilon > 0$ such that $PM(k, 1/2 - \varepsilon)$ is in P. This leaves a hardness gap for $PM(k, \delta)$ when $\delta \in [1/k, 1/2 - \varepsilon]$.

The connection with the work of this paper is that we can check in polynomial time whether or not the edges of a k-graph H satisfy arithmetic conditions of the types which define our notion of a divisibility barrier. We will see that for $\delta > 1/k$, any sufficiently large k-graph H with $\delta(H) \geq \delta|V(H)|$ cannot be close to a space barrier, so our main theorem will imply that either H contains a perfect matching or H is close to a divisibility barrier. So to decide $PM(k, \delta)$ for $\delta > 1/k$ it suffices to decide the existence in a perfect matching when H is close to a divisibility barrier. Unfortunately, in our main theorem our notion of 'close' means an edit-distance of $o(n^k)$, which is too large to be checked by a brute-force approach. However, for k-graphs H of codegree close to $n/2$ we are able to refine our main theorem to prove the following result, which states that either H admits a perfect matching or *every* edge of H satisfies a divisibility condition of a type which defines a divisibility barrier.

THEOREM 1.3. *For any $k \geq 3$ there exists $c > 0$ and n_0 such that for any $n \geq n_0$ with $k \mid n$ and any k-graph H on n vertices with $\delta(H) \geq (1/2 - c)n$ the following statement holds. H does not contain a perfect matching if and only if there exists a partition of $V(H)$ into parts V_1, V_2 of size at least $\delta(G)$ and $a \in \{0, 1\}$ so that $|V_1| \neq an/k \mod 2$ and $|e \cap V_1| = a \mod 2$ for all edges e of G.*

It is not hard to construct an algorithm with polynomial running time which checks the existence of such a partition, and so we recover the aforementioned result of Karpiński, Ruciński and Szymańska that $PM(k, 1/2 - c)$ is in P. Together with Knox [**24**], we were able to refine the methods of this paper to prove stronger results for k-graphs of large minimum degree to almost eliminate the hardness gap referred to above. Indeed, we show that we may replace the condition $\delta(H) \geq (1/2 - c)n$ in Theorem 1.3 by the condition $\delta(H) \geq n/3 + o(n)$, which shows that $\mathrm{PM}(k, \delta)$ is in P for any $\delta > 1/3$. Furthermore, we are able to then extend Theorem 1.3 by further reducing the bound on $\delta(H)$; although this requires a significantly more complicated statement, we obtain a polynomial-time algorithm which decides $\mathrm{PM}(k, \delta)$ for $\delta > 1/k$. Together with the work of Szymańska described above, this settles the complexity status of $\mathrm{PM}(k, \delta)$ for any $\delta \neq 1/k$. However, these refinements of our results only apply to k-graphs of large minimum codegree, whilst the results of this paper are much more general.

1.5. Notation

The following notation is used throughout the paper: $[k] = \{1, \ldots, k\}$; if X is a set then $\binom{X}{k}$ is the set of subsets of X of size k, and $\binom{X}{\leq k} = \bigcup_{i \leq k} \binom{X}{i}$ is the set of subsets of X of size at most k; $o(1)$ denotes a term which tends to zero as n tends to infinity; $x \ll y$ means that for every $y > 0$ there exists some $x_0 > 0$ such that the subsequent statement holds for any $x < x_0$ (such statements with more variables are defined similarly). We write $x = y \pm z$ to mean $y - z \leq x \leq y + z$. Also we denote all vectors in bold font, and their coordinates by subscripts of the same character in standard font, e.g. $\mathbf{a} = (a_1, \ldots, a_n)$.

CHAPTER 2

Results and examples

In this chapter we state our main theorems on perfect matchings in simplicial complexes. For almost perfect matchings it requires no additional work to obtain more general results that dispense with the 'downward closure' assumption. However, for perfect matchings it is more convenient to assume downward closure, which seems to hold in any natural application of our results, so we will stick to simplicial complexes, and make some remarks later on how the approach may be generalised. We also discuss several examples that illustrate various aspects of the theory: space barriers and tightness of the degree condition, lattice-based constructions, and generalisations of previous results.

2.1. Almost perfect matchings

We start with some definitions. We identify a hypergraph H with its edge set, writing $e \in H$ for $e \in E(H)$, and $|H|$ for $|E(H)|$. A k-*system* is a hypergraph J in which every edge of J has size at most k and $\emptyset \in J$. We refer to the edges of size r in J as r-*edges of* J, and write J_r to denote the r-graph on $V(J)$ formed by these edges. It may be helpful to think of the r-graphs J_r as different 'levels' of the k-system J. A k-*complex* J is a k-system whose edge set is closed under inclusion, i.e. if $e \in H$ and $e' \subseteq e$ then $e' \in H$. That is, each level of J is supported by the levels below it. For any non-empty k-graph G, we may generate a k-complex $G^{\leq}$ whose edges are any $e \subseteq V(G)$ such that $e \subseteq e'$ for some edge $e' \in G$.

We introduce the following notion of degree in a k-system J. For any edge e of J, the *degree* $d(e)$ of e is the number of $(|e|+1)$-edges e' of J which contain e as a subset. (Note that this is *not* the standard notion of degree used in k-graphs, in which the degree of a set is the number of edges containing it.) The *minimum r-degree* of J, denoted $\delta_r(J)$, is the minimum of $d(e)$ taken over all r-edges $e \in J$. So every r-edge of J is contained in at least $\delta_r(J)$ of the $(r+1)$-edges of J. Note that if J is a k-complex then $\delta_r(J) \leq \delta_{r-1}(J)$ for each $r \in [k-1]$. The *degree sequence of* J is

$$\delta(J) = (\delta_0(J), \delta_1(J), \ldots, \delta_{k-1}(J)).$$

Our minimum degree assumptions will always be of the form $\delta(J) \geq \mathbf{a}$ pointwise for some vector $\mathbf{a} = (a_0, \ldots, a_{k-1})$, i.e. $\delta_i(J) \geq a_i$ for $0 \leq i \leq k-1$. It is helpful to interpret this 'dynamically' as follows: when constructing an edge of J_k by greedily choosing one vertex at a time, there are at least a_i choices for the $(i+1)$st vertex (this is the reason for the requirement that $\emptyset \in J$, which we need for the first choice in the process). To see that this generalises the setting of minimum degree in hypergraphs, consider any k-graph G and let J be the k-system on $V(G)$ with $J_k = G$ and complete lower levels $J_i = \binom{V(G)}{i}$, $0 \leq i \leq k-1$. Then $\delta(J) = (n, n-1, \ldots, n-k+2, \delta(G))$ is a degree sequence in which the first

7

$k-1$ coordinates are as large as possible, so any minimum degree condition for this k-system J reduces to a minimum degree condition for G. More generally, for any $i \geq 0$, define an *i-clique* of G to be a set $I \subseteq V(G)$ of size i such that every $K \subseteq I$ of size k is an edge of G (this last condition is vacuous if $i < k$). We can then naturally define a *clique r-complex* of G, whose edges are the cliques in G.

DEFINITION 2.1 (Clique r-complex). *The* clique r-complex $J_r(G)$ *of a k-graph G is defined by taking $J_r(G)_i$ to consist of the i-cliques of G for $0 \leq i \leq r$.*

To see the power in the extra generality of the degree sequence setting, consider the 4-graph of tetrahedra in a 3-graph G. This does not satisfy any minimum degree condition as a 4-graph, as non-edges of G are not contained in any 4-edges; on the other hand, we will see later that a minimum degree condition on G implies a useful minimum degree condition for the clique 4-complex.

Our first theorem on hypergraph matchings gives a sufficient minimum degree sequence for a k-system to contain a matching covering almost all of its vertices.

THEOREM 2.2. *Suppose that $1/n \ll \alpha \ll 1/k$, and that J is a k-system on n vertices with*

$$\delta(J) \geq \left(n, \left(\frac{k-1}{k} - \alpha\right)n, \left(\frac{k-2}{k} - \alpha\right)n, \ldots, \left(\frac{1}{k} - \alpha\right)n\right).$$

Then J_k contains a matching which covers all but at most $9k^2\alpha n$ vertices of J.

Next we describe a family of examples showing that Theorem 2.2 is best possible up to the constant $9k^2$ (which we do not attempt to optimise), in the sense that there may not be any almost perfect matching if *any* coordinate of the degree sequence is substantially reduced.

CONSTRUCTION 2.3. *(Space barriers) Suppose V is a set of n vertices, $j \in [k-1]$ and $S \subseteq V$. Let $J = J(S,j)$ be the k-complex in which J_i (for $0 \leq i \leq k$) consists of all i-sets in V that contain at most j vertices of S.*

Figure 1 illustrates this construction in the case $k = 3$ and $j = 2$, along with the corresponding *transferral digraph*, which will be defined formally later. Observe that for any $i \leq j$ the i-graph $J_i = \binom{V}{i}$ is complete, so we have $\delta_i(J) = n - i$ for $0 \leq i \leq j-1$ and $\delta_i(J) = n - |S| - (i-j)$ for $i \geq j$. Choosing $|S| = \lfloor(\frac{j}{k} + \alpha)n\rfloor - k$, we see that the minimum degree condition of Theorem 2.2 is satisfied. However, every edge of J_k has at least $k - j$ vertices in $V \setminus S$, so the maximum matching in J_k has size $\lfloor \frac{|V \setminus S|}{k-j} \rfloor$, which leaves at least $\alpha n - k$ uncovered vertices.

In these examples we have a space barrier for matchings, which is robust in the sense that the number of uncovered vertices varies smoothly with the minimum degree sequence. Our preliminary stability result is that space barriers are the only obstructions to finding a matching that covers all but a constant number of vertices. Note that we cannot avoid leaving a constant number of uncovered vertices because of divisibility barriers (discussed below). Our structures permit some small imperfections, defined as follows. Suppose G and H are k-graphs on the same set V of n vertices and $0 < \beta < 1$. We say that G is β *contained* in H if all but at most βn^k edges of G are edges of H.

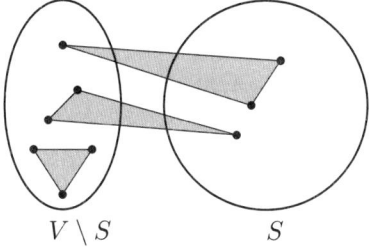
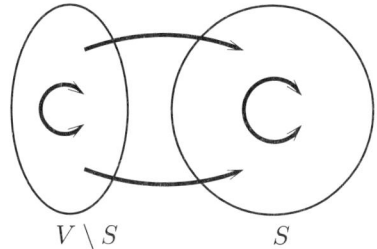

FIGURE 1. The left hand diagram shows an example of a space barrier H: the edges are all 3-tuples which have at most two vertices in S. If $|S| > \frac{2}{3}|V(H)|$ then H has no perfect matching. If instead $|S| = \frac{2}{3}|V(H)|$ then a perfect matching M in H must consist only of edges with exactly two vertices in S, whereupon for any ℓ the transferral digraph $D_\ell(H, M)$ is as shown in the right hand diagram; it contains all edges except those directed from S to $V \setminus S$.

THEOREM 2.4. *Suppose that $1/n \ll 1/\ell \ll \alpha \ll \beta \ll 1/k$. Let J be a k-complex on n vertices such that*

$$\delta(J) \geq \left(n, \left(\frac{k-1}{k} - \alpha\right)n, \left(\frac{k-2}{k} - \alpha\right)n, \ldots, \left(\frac{1}{k} - \alpha\right)n\right).$$

Then J has at least one of the following properties:

1 (Matching): J_k *contains a matching that covers all but at most ℓ vertices.*
2 (Space barrier): J_k *is β-contained in $J(S, j)_k$ for some $j \in [k-1]$ and $S \subseteq V(J)$ with $|S| = \lfloor jn/k \rfloor$.*

2.2. Partite systems

We will also require 'partite' analogues of our hypergraph matching theorems. A *partition* $\mathcal{P}$ of a set V is a sequence of disjoint sets $V_1, \ldots, V_k$ whose union is V; we refer to the sets V_i as the *parts* of $\mathcal{P}$, and write $U \in \mathcal{P}$ to mean that U is a part of $\mathcal{P}$. Note that we consider the partition $\mathcal{P}$ to describe not just the contents of each part but also the order of the parts. We say that the partition $\mathcal{P}$ is *balanced* if each part has the same size.

Let H be a hypergraph, and let $\mathcal{P}$ be a partition of $V(H)$. Then we say a set $S \subseteq V(H)$ is $\mathcal{P}$-*partite* if if has at most one vertex in any part of $\mathcal{P}$, and that H is $\mathcal{P}$-*partite* if every edge of H is $\mathcal{P}$-partite. We say that H is r-*partite* if there exists some partition $\mathcal{P}$ of $V(H)$ into r parts such that H is $\mathcal{P}$-partite. For r-partite k-systems we introduce the following alternative notion of degree. Let V be a set of vertices, let $\mathcal{P}$ be a partition of V into r parts $V_1, \ldots, V_r$, and let J be a $\mathcal{P}$-partite k-system on V. For each $0 \leq j \leq k-1$ we define the *partite minimum j-degree* $\delta_j^*(J)$ as the largest m such that any j-edge e has at least m extensions to a $(j+1)$-edge in any part not used by e, i.e.

$$\delta_j^*(J) := \min_{e \in J_j} \min_{i: e \cap V_i = \emptyset} |\{v \in V_i : e \cup \{v\} \in J\}|.$$

The *partite degree sequence* is $\delta^*(J) = (\delta_0^*(J), \ldots, \delta_{k-1}^*(J))$. Note that we suppress the dependence on $\mathcal{P}$ in our notation: this will be clear from the context.

Our next theorem is an analogue of Theorem 2.2 for r-partite k-systems. Here we may add an additional condition on our matching, for which we need the following definition. Suppose that $\mathcal{P}$ is a partition of a set of vertices V into vertex classes $V_1, \ldots, V_r$, and H is a $\mathcal{P}$-partite k-graph on V. Then the *index set* of an edge $e \in H$ is $\{i : |e \cap V_i| = 1\} \in \binom{[r]}{k}$. For a matching M in H and a set $A \in \binom{[r]}{k}$ let $N_A(M)$ denote the number of edges $e \in M$ whose index set is A. We say that M is *balanced* if $N_A(M)$ is constant over all $A \in \binom{[r]}{k}$, that is, each index set is represented by equally many edges. The following theorem shows that we may insist that the matching obtained is balanced.

THEOREM 2.5. *Suppose that $1/n \ll \alpha \ll 1/r \leq 1/k$, and that J is a r-partite k-system on vertex classes each of n vertices with*

$$\delta^*(J) \geq \left(n, \left(\frac{k-1}{k} - \alpha\right) n, \left(\frac{k-2}{k} - \alpha\right) n, \ldots, \left(\frac{1}{k} - \alpha\right) n\right).$$

Then J_k contains a balanced matching which covers all but at most $9k^2 r\alpha n$ vertices of J.

To see that this is best possible in the same sense as for Theorem 2.2 we use the natural partite version of Construction 2.3, which shows that there may not be any almost perfect matching if any coordinate of the partite degree sequence is substantially reduced.

CONSTRUCTION 2.6. *(Partite space barriers)* Suppose $\mathcal{P}$ partitions a set V into r parts $V_1, \ldots, V_r$ of size n. Let $j \in [k-1]$, $S \subseteq V$ and $J = J_r(S, j)$ be the k-complex in which J_i (for $0 \leq i \leq k$) consists of all $\mathcal{P}$-partite i-sets in V that contain at most j vertices of S.

We choose S to have $s = \lfloor (\frac{j}{k} + \alpha) n \rfloor$ vertices in each part. Then $\delta_i^*(J) = n$ for $0 \leq i \leq j-1$ and $\delta_i^*(J) = n - s$ for $j \leq i \leq k-1$, so the minimum partite degree condition of Theorem 2.5 is satisfied. However, the maximum matching in J_k has size $\lfloor \frac{|V \setminus S|}{k-j} \rfloor$, which leaves at least $r(\alpha n - k)$ uncovered vertices.

We also have the following stability result analogous to Theorem 2.4.

THEOREM 2.7. *Suppose that $1/n \ll 1/\ell \ll \alpha \ll \beta \ll 1/r \leq 1/k$. Let $\mathcal{P}$ partition a set V into parts $V_1, \ldots, V_r$ each of size n, and let J be a $\mathcal{P}$-partite k-complex on V with*

$$\delta^*(J) \geq \left(n, \left(\frac{k-1}{k} - \alpha\right) n, \left(\frac{k-2}{k} - \alpha\right) n, \ldots, \left(\frac{1}{k} - \alpha\right) n\right).$$

Then J has at least one of the following properties:

1 (Matching): *J_k contains a matching that covers all but at most ℓ vertices.*
2 (Space barrier): *J is β-contained in $J_r(S,j)_k$ for some $j \in [k-1]$ and $S \subseteq V$ with $\lfloor jn/k \rfloor$ vertices in each V_i, $i \in [r]$.*

2.3. Lattice-based constructions

Having described the general form of space barriers, we now turn our attention to divisibility barriers. For this we need the notion of *index vectors*, which will play a substantial role in this paper. Let V be a set of vertices, and let $\mathcal{P}$ be a partition

of V into d parts $V_1, \ldots, V_d$. Then for any $S \subseteq V$, the *index vector of S with respect to $\mathcal{P}$* is the vector
$$\mathbf{i}_\mathcal{P}(S) := (|S \cap V_1|, \ldots, |S \cap V_d|) \in \mathbb{Z}^d.$$
When $\mathcal{P}$ is clear from the context, we write simply $\mathbf{i}(S)$ for $\mathbf{i}_\mathcal{P}(S)$. So $\mathbf{i}(S)$ records how many vertices of S are in each part of $\mathcal{P}$. Note that $\mathbf{i}(S)$ is well-defined as we consider the partition $\mathcal{P}$ to define the order of its parts.

CONSTRUCTION 2.8. *(Divisibility barriers) Suppose L is a lattice in $\mathbb{Z}^d$ (i.e. an additive subgroup) with $\mathbf{i}(V) \notin L$, fix any $k \geq 2$, and let G be the k-graph on V whose edges are all k-tuples e with $\mathbf{i}(e) \in L$.*

For any matching M in G with vertex set $S = \bigcup_{e \in M} e$ we have $\mathbf{i}(S) = \sum_{e \in M} \mathbf{i}(e) \in L$. Since we assumed that $\mathbf{i}(V) \notin L$ it follows that G does not have a perfect matching.

We will now give some concrete illustrations of this construction.

(1) Suppose $d = 2$ and $L = \langle(-2, 2), (0, 1)\rangle$. Note that $(x, y) \in L$ precisely when x is even. Then by definition, $|V_1|$ is odd, and the edges of G are all k-tuples $e \subseteq V$ such that $|e \cap V_1|$ is even. If $|V| = n$ and $|V_1| \sim n/2$, then we recover the extremal example mentioned earlier for the minimum degree perfect matching problem. The generated k-complex $G^\leq$ has $\delta_i(G^\leq) \sim n$ for $0 \leq i \leq k - 2$ and $\delta_{k-1}(G^\leq) \sim n/2$.

(2) Suppose $d = 3$ and $L = \langle(-2, 1, 1), (1, -2, 1), (1, 0, 0)\rangle$. Note that $(x, y, z) \in L$ precisely when $y = z$ mod 3 (This construction is illustrated in Figure 2, again with the accompanying transferral digraph, which will be defined formally later.). Thus $|V_2| \neq |V_3|$ mod 3 and the edges of G are all k-tuples $e \subseteq V$ such that $|e \cap V_2| = |e \cap V_3|$ mod 3. Note that for any $(k-1)$-tuple $e' \subseteq V$ there is a unique $j \in [3]$ such that such that $\mathbf{i}(e') + \mathbf{u}_j \in L$, where $\mathbf{u}_j$ is the jth standard basis vector. (We have $j = 1$ if $\mathbf{i}(e')_2 = \mathbf{i}(e')_3$ mod 3, $j = 2$ if $\mathbf{i}(e')_2 = \mathbf{i}(e')_3 - 1$ mod 3, or $j = 3$ if $\mathbf{i}(e')_2 = \mathbf{i}(e')_3 + 1$ mod 3.) If $|V| = n$ and $|V_1|, |V_2|, |V_3| \sim n/3$ then $\delta_i(G^\leq) \sim n$ for $0 \leq i \leq k - 2$ and $\delta_{k-1}(G^\leq) \sim n/3$. Mycroft [42] showed that this construction is asymptotically extremal for a range of hypergraph packing problems. For example, he showed that any 4-graph G on n vertices with $7 \mid n$ and $\delta(G) \geq n/3 + o(n)$ contains a perfect $K_{4,1,1,1}^4$-packing, where $K_{4,1,1,1}^4$ denotes the complete 4-partite 4-graph with vertex classes of size $4, 1, 1$ and 1 respectively; this construction demonstrates that this bound is best possible up to the $o(n)$ error term.

For simplicity we only gave approximate formulae for the degree sequences in these examples, but it is not hard to calculate them exactly. Note that the index vectors of edges in a k-graph belong to the hyperplane $\{\mathbf{x} \in \mathbb{R}^d : \sum_{i \in [d]} x_i = k\}$. The intersection of this hyperplane with the lattice L is a coset of the intersection of L with the hyperplane $\Pi^d = \{\mathbf{x} \in \mathbb{R}^d : \sum_{i \in [d]} x_i = 0\}$. Thus it is most convenient to choose a basis of L that includes a basis of $L \cap \Pi^d$, as in the above two examples. In order for the construction to exist it must be possible to satisfy $\mathbf{i}(V) \notin L$, so L should not contain the lattice $M_k^d = \{\mathbf{x} \in \mathbb{Z}^d : k \mid \sum_{i \in [d]} x_i\}$. We say that L is *complete* if $M_k^d \subseteq L$, otherwise L is *incomplete*. Assuming that L contains some $\mathbf{v}$ with $\sum v_i = k$, an equivalent formulation of completeness is that $L \cap \Pi^d = \mathbb{Z}^d \cap \Pi^d$, since this is equivalent to $L \cap (\Pi^d + \mathbf{v}) = \mathbb{Z}^d \cap (\Pi^d + \mathbf{v}) = \{\mathbf{x} \in \mathbb{Z}^d : \sum_{i \in [d]} x_i = k\}$, and so to $M_k^d \subseteq L$.

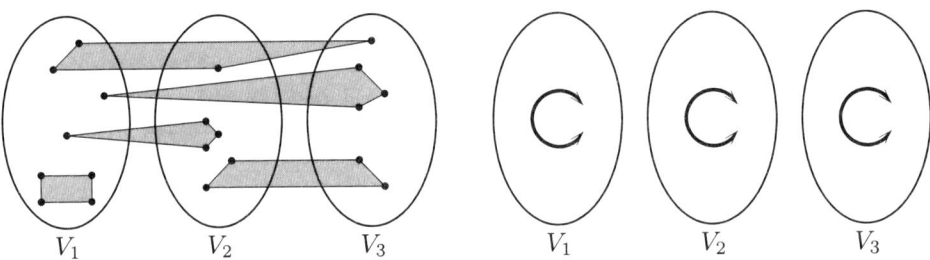

FIGURE 2. The left hand diagram shows an example of a divisibility barrier H: the edges are all 4-tuples e with $|e \cap V_2| \equiv |e \cap V_3|$ mod 3. So if $|V_2| \neq |V_3|$ mod 3 then there is no perfect matching in H. If instead $|V_2| = |V_3|$ and H contains a perfect matching, then for any ℓ the transferral digraph $D_\ell(H, M)$ is as shown in the right hand diagram; it consists of three disjoint cliques induced by V_1, V_2 and V_3.

There is a natural notion of minimality for lattice-based constructions. We say that L is *transferral-free* if it does not contain any difference of standard basis vectors, i.e. a vector of the form $\mathbf{u}_i - \mathbf{u}_j$ with $i \neq j$. If L is not transferral-free, then the construction for L can be reformulated using a lattice of smaller dimension. Without loss of generality we may consider the case that $\mathbf{u}_{d-1} - \mathbf{u}_d \in L$. Then we replace $\mathcal{P}$ by the partition of V into $d-1$ parts obtained by combining V_{d-1} and V_d, and replace L by the lattice L' consisting of all $\mathbf{x} \in \mathbb{Z}^{d-1}$ such that $(x_1, \ldots, x_{d-1}, 0) \in L$. We remark that a transferral-free lattice in $\mathbb{Z}^d$ has index at least d as a subgroup of $\mathbb{Z}^d$. To see this, note that the cosets $L + (\mathbf{u}_1 - \mathbf{u}_i)$, $1 \leq i \leq d$ must be distinct, otherwise there is some $\mathbf{x} \in Z^d$ and $i, j \in [d]$ for which $\mathbf{x} = \mathbf{v} + \mathbf{u}_1 - \mathbf{u}_i = \mathbf{v}' + \mathbf{u}_1 - \mathbf{u}_j$ with $\mathbf{v}, \mathbf{v}' \in L$, which gives $\mathbf{u}_i - \mathbf{u}_j = \mathbf{v} - \mathbf{v}' \in L$, contradicting the fact that L is transferral-free.

2.4. Perfect matchings

For our main theorems we specialise to the setting of simplicial complexes, which will simplify some arguments and is sufficiently general for our applications. Under the same minimum degree assumption that is tight for almost perfect matchings, we refine the results above, by showing that either we have a perfect matching, or we have a structural description for the complex: it either has a space barrier or a divisibility barrier. It is remarkable that this rigidity emerges from the purely combinatorial degree sequence assumption, and that it exhibits these two very different phenomena, one tied to convex geometry and the other to integer lattices.

Our structures permit some small imperfections, defined as follows. Recall that we say that a k-graph G is β-contained in a k-graph H if all but at most βn^k edges of G are edges of H. Also, given a partition $\mathcal{P}$ of V into d parts, we define the μ-robust edge lattice $L_\mathcal{P}^\mu(G) \subseteq \mathbb{Z}^d$ to be the lattice generated by all vectors $\mathbf{v} \in \mathbb{Z}^d$ such that there are at least μn^k edges $e \in G$ with $\mathbf{i}_\mathcal{P}(e) = \mathbf{v}$. Recall that we call a lattice $L \subseteq \mathbb{Z}^d$ complete if $L \cap \Pi^d = \mathbb{Z}^d \cap \Pi^d$, where $\Pi^d = \{\mathbf{x} \in \mathbb{R}^d : \sum_{i \in [d]} x_i = 0\}$. Recall also that the space barrier constructions $J(S, j)$ were defined in Section 2.1. Now we can state our first main theorem.

THEOREM 2.9. *Suppose that $1/n \ll \alpha \ll \mu, \beta \ll 1/k$ and that $k \mid n$. Let J be a k-complex on n vertices such that*

$$\delta(J) \geq \left(n, \left(\frac{k-1}{k} - \alpha\right)n, \left(\frac{k-2}{k} - \alpha\right)n, \ldots, \left(\frac{1}{k} - \alpha\right)n\right).$$

Then J has at least one of the following properties:

1 **(Matching)**: *J_k contains a perfect matching.*
2 **(Space barrier)**: *J_k is β-contained in $J(S,j)_k$ for some $j \in [k-1]$ and $S \subseteq V(J)$ with $|S| = jn/k$.*
3 **(Divisibility barrier)**: *There is some partition $\mathcal{P}$ of $V(J)$ into $d \leq k$ parts of size at least $\delta_{k-1}(J) - \mu n$ such that $L_\mathcal{P}^\mu(J_k)$ is incomplete and transferral-free.*

Our second main theorem is a partite version of the previous result. Recall that the space barrier constructions in this setting are described in Construction 2.6. We also need to account for the original partition when classifying edge lattices as follows. Suppose $\mathcal{P}$ is a partition of V into d parts $(V_1, \ldots, V_d)$ that refines a partition $\mathcal{P}'$ of V. Let $L_{\mathcal{P}\mathcal{P}'} \subseteq \mathbb{Z}^d \cap \Pi^d$ be the lattice generated by all differences of basis vectors $\mathbf{u}_i - \mathbf{u}_j$ for which V_i, V_j are contained in the same part of $\mathcal{P}'$. We say that a lattice $L \subseteq \mathbb{Z}^d$ is *complete with respect to $\mathcal{P}'$* if $L_{\mathcal{P}\mathcal{P}'} \subseteq L \cap \Pi^d$, otherwise we say that L is *incomplete with respect to $\mathcal{P}'$*.

Similarly to Theorem 2.9, this theorem tells us that an r-partite k-complex satisfying the given minimum degree condition either contains a perfect matching or has a space barrier or divisibility barrier. However, in applications (for example the multipartite Hajnal-Szemerédi theorem proved in [**26**]) will may need to know that the perfect matching obtained has roughly the same number of edges of each index. Recall that the index set of an edge $e \in H$ is $\{i : |e \cap V_i| = 1\} \in \binom{[r]}{k}$, that for a matching M in H we write $N_A(M)$ to denote the number of edges $e \in M$ whose index set is A, and that M is balanced if $N_A(M)$ is constant over all $A \in \binom{[r]}{k}$. Unfortunately, Construction 2.11 will show that we cannot insist on a balanced matching in our partite analogue of Theorem 2.9. Instead we require a weaker property: we say that M is γ-*balanced* if $N_A(M) \geq (1-\gamma)N_B(M)$ for any $A, B \in \binom{[r]}{k}$, meaning that M is close to being balanced.

THEOREM 2.10. *Suppose that $1/n \ll \gamma, \alpha \ll \mu, \beta \ll 1/r \leq 1/k$. Let $\mathcal{P}'$ partition a set V into parts $V_1, \ldots, V_r$ each of size n, where $k \mid rn$. Suppose that J is a $\mathcal{P}'$-partite k-complex with*

$$\delta^*(J) \geq \left(n, \left(\frac{k-1}{k} - \alpha\right)n, \left(\frac{k-2}{k} - \alpha\right)n, \ldots, \left(\frac{1}{k} - \alpha\right)n\right).$$

Then J has at least one of the following properties:

1 **(Matching)**: *J_k contains a γ-balanced perfect matching.*
2 **(Space barrier)**: *J_k is β-contained in $J_r(S,j)_k$ for some $j \in [k-1]$ and $S \subseteq V$ with $\lfloor jn/k \rfloor$ vertices in each V_i, $i \in [r]$.*
3 **(Divisibility barrier)**: *There is some partition $\mathcal{P}$ of $V(J)$ into $d \leq kr$ parts of size at least $\delta_{k-1}^*(J) - \mu n$ such that $\mathcal{P}$ refines $\mathcal{P}'$ and $L_\mathcal{P}^\mu(J_k)$ is incomplete with respect to $\mathcal{P}'$ and transferral-free.*

A necessary condition for the existence of a balanced perfect matching in an r-partite k-graph whose vertex classes have size n is that rn/k, the number of edges, is divisible by $\binom{r}{k}$, the number of possible index sets of edges. However, even under this additional assumption we cannot replace 'γ-balanced' with 'balanced' in option 1 of this theorem; some small imbalance may be inevitable, as shown by the following construction.

CONSTRUCTION 2.11. *Choose integers $r \geq 3$ and n so that $2n/(r-1)$ and rn are integers of different parity (e.g. $r = 5$ and $n \equiv 2 \mod 4$). Let G be an r-partite graph whose vertex classes each have size two, say $\{x_i, y_i\}$ for $i \in [r]$, and whose edges are $x_1 y_2$, $x_2 y_1$, and $x_i x_j$ and $y_i y_j$ for any pair $\{i, j\}$ other than $\{1, 2\}$. Form the 'blow-up' G^* by replacing every vertex of G with n vertices (for some even integer n), and adding edges between any pair of vertices which replace adjacent vertices in G.*

Observe that the graph G^* constructed in Construction 2.11 is an r-partite graph whose vertex classes each have size $2n$ and which satisfies $\delta^*(G^*) = n$. So the 2-complex $J = (G^*)^{\leq}$ satisfies the conditions of Theorem 2.10. Furthermore, it is not hard to check that J does not satisfy options 2 or 3 of Theorem 2.10 for small μ, β. So $J_2 = G^*$ must contain a perfect matching which is close to being balanced, and indeed it is not hard to verify this. However, there is no balanced perfect matching in G^*. Indeed, let Y be the set of vertices of G^* which replaced one of the vertices y_i of G, so $|Y| = rn$. In a balanced perfect matching M each index set would be represented by $rn/\binom{r}{2} = \frac{2n}{r-1}$ edges of M. This means that the number of vertices of Y covered by edges of M of index $\{1,2\}$ is $\frac{2n}{r-1}$, and so $|Y| - \frac{2n}{r-1} = rn - \frac{2n}{r-1}$ must be even, a contradiction.

Theorems 2.9 and 2.10 will each be deduced from a more general statement, Theorem 7.11, in Chapter 7. However, given a slightly stronger degree sequence, we can rule out the possibility of a space barrier in each of these theorems. This will frequently be the case in applications, so for ease of use we now give the corresponding theorems.

THEOREM 2.12. *Suppose that $1/n \ll \mu \ll \alpha, 1/k$ and that $k \mid n$. Let J be a k-complex on n vertices such that*

$$\delta(J) \geq \left(n, \left(\frac{k-1}{k} + \alpha\right)n, \left(\frac{k-2}{k} + \alpha\right)n, \ldots, \left(\frac{1}{k} + \alpha\right)n\right).$$

Then J has at least one of the following properties:
  1 (Matching): *J_k contains a perfect matching.*
  2 (Divisibility barrier): *There is some partition $\mathcal{P}$ of $V(J)$ into $d \leq k$ parts of size at least $\delta_{k-1}(J) - \mu n$ such that $L^{\mu}_{\mathcal{P}}(J_k)$ is incomplete and transferral-free.*

PROOF. Introduce a new constant β with $1/n \ll \beta \ll \alpha, 1/k$. Then it suffices to show that option 2 of Theorem 2.9 is impossible. So suppose for a contradiction that J_k is β-contained in $J(S, j)_k$ for some $j \in [k-1]$ and $S \subseteq V(J)$ with $|S| = jn/k$. We now form an ordered k-tuple $(x_1, \ldots, x_k)$ of vertices of J such that $\{x_1, \ldots x_s\} \in J$ for any $s \in [k]$ and $x_1, \ldots, x_{j+1} \in S$. Indeed, the minimum degree sequence ensures that when greedily choosing the vertices one at a time there are at least $n - |S| \geq n/k$ choices for the first vertex, at least $\delta_i(J) - (n - |S|) \geq (j - i)n/k + \alpha n$ choices for the $(i + 1)$st vertex for any $i \leq j$, and at least $\delta_i(J) \geq (k - i)n/k + \alpha n$ for the $(i + 1)$st vertex for any $i > j$. In total this gives at least $\alpha(n/k)^k$ ordered k-tuples, each of which is an edge of J_k with at least $j+1$ vertices in S, so is not contained in $J(S,j)_k$. This is a contradiction for $k!\beta < \alpha/k^k$. $\square$

THEOREM 2.13. *Suppose that $1/n \ll \gamma, \mu \ll \alpha, 1/r, 1/k$ and $r \geq k$. Let $\mathcal{P}'$ partition a set V into parts $V_1, \ldots, V_r$ each of size n, where $k \mid rn$. Suppose that J is a $\mathcal{P}'$-partite k-complex with*

$$\delta^*(J) \geq \left(n, \left(\frac{k-1}{k} + \alpha\right)n, \left(\frac{k-2}{k} + \alpha\right)n, \ldots, \left(\frac{1}{k} + \alpha\right)n\right).$$

Then J has at least one of the following properties:
  1 **(Matching):** J_k *contains a γ-balanced perfect matching.*
  2 **(Divisibility barrier):** *There is some partition $\mathcal{P}$ of $V(J)$ into $d \leq kr$ parts of size at least $\delta^*_{k-1}(J) - \mu n$ such that $\mathcal{P}$ refines $\mathcal{P}'$ and $L^\mu_\mathcal{P}(J_k)$ is incomplete with respect to $\mathcal{P}'$ and transferral-free.*

This theorem follows from Theorem 2.10 exactly as Theorem 2.12 followed from Theorem 2.9; we omit the details.

2.5. Further results

Theorems 2.9 and 2.10 can be applied to a variety of matching and packing problems in graphs and hypergraphs. Indeed, in this chapter we give a short deduction of Theorem 1.2 from Theorem 2.13 (which was a consequence of Theorem 2.10), whilst in Chapter 8 we use Theorem 2.9 to prove Theorem 1.1. We can also recover and find new variants of existing results. For example, consider the result of Rödl, Ruciński and Szemerédi [49] on the minimum degree threshold for a perfect matching in a k-graph. Their proof proceeds by a stability argument, giving a direct argument when the k-graph is close to an extremal configuration, and otherwise showing that even a slightly lower minimum degree is sufficient for a perfect matching. Our results give a new proof of stability under a much weaker degree assumption. In the following result, we only assume that the minimum degree of G is a bit more than $n/3$, and show that if there is no perfect matching then G is almost contained in an extremal example. As described earlier, with a bit more work (and a stronger degree assumption) we can show that G is contained in an extremal example (see Theorem 1.3). This requires some technical preliminaries, so we postpone the proof for now.

THEOREM 2.14. *Suppose $1/n \ll b, c \ll 1/k$, $k \geq 3$, k divides n and G is a k-graph on n vertices with $\delta(G) \geq (1/3 + c)n$ and no perfect matching. Then there is a partition of $V(G)$ into parts V_1, V_2 of size at least $\delta(G)$ and $a \in \{0, 1\}$ so that all but at most bn^k edges e of G have $|e \cap V_1| = a \bmod 2$.*

PROOF. Introduce a constant μ with $1/n \ll \mu \ll b, c$. Let J be the (clique) k-complex with $J_k = G$ and J_i complete for $i < k$. Suppose that J_k does not have a perfect matching. Then option 2 must hold in Theorem 2.12, that is, there is some partition $\mathcal{P}$ of $V(J)$ into parts of size at least $\delta(G) - \mu n$ such that $L^\mu_\mathcal{P}(J_k)$ is incomplete. Since $\delta(G) - \mu n > n/3$ there must be 2 such parts, and $L^\mu_\mathcal{P}(J_k) \cap \Pi^2$ is generated by $(-t, t)$ for some $t \geq 2$. We cannot have $t > 2$, as then neither $(k-2, 2)$ nor $(k-1, 1)$ lie in $L^\mu_\mathcal{P}(J_k)$, and so there are at most $2\mu n^k$ edges of $J_k = G$ with one of these two indices. Since there are at least $(n/3)^{k-1}/(k-1)!$ $(k-1)$-tuples of index $(k-2, 1)$, by averaging some such $(k-1)$-tuple must be contained in at most $3^k k! \mu n^k$ edges of G, contradicting the minimum degree assumption. So we must have $t = 2$, and the result follows. □

We now present the deduction of Theorem 1.2 from Theorem 2.13.

Proof of Theorem 1.2. Introduce a constant μ with $1/n \ll \mu \ll c$. Let J be the clique k-complex of G. Then $\delta_i(J) \geq (k-i)n/k + icn$ for $0 \leq i \leq k-1$. Thus we can apply Theorem 2.13. Suppose for a contradiction that option 2 of this theorem holds. Then there is some partition $\mathcal{P}$ of $V(J)$ into parts of size at least $\delta_{k-1}^*(J) - \mu n$ such that $\mathcal{P}$ refines the partition $\mathcal{P}'$ of V into $V_1, \cdots, V_k$ and $L_\mathcal{P}^\mu(J_k)$ is incomplete with respect to $\mathcal{P}'$. We may assume that $L_\mathcal{P}^\mu(J_k)$ is transferral-free; recall that this means that it does not contain any difference of standard basis vectors $\mathbf{u}_i - \mathbf{u}_j$ with $i \neq j$. Consider a part of $\mathcal{P}'$ that is refined in $\mathcal{P}$, without loss of generality it is V_1, and vertices $x_1, x_1' \in V_1$ in different parts U_1, U_1' of $\mathcal{P}$. We can greedily construct many sequences $x_i \in V_i$, $2 \leq i \leq k$ such that $x_1 x_2 \ldots x_k$ and $x_1' x_2 \ldots x_k$ are both k-cliques: since $\delta^*(G) \geq (k-1)n/k + cn$ there are at least $(k-i)n/k + icn$ choices for x_i, so at least $c(n/k)^{k-1}$ such sequences. We can repeat this for all choices of $x_1 \in U_1$ and $x_1' \in U_1'$; there are at least $\delta_{k-1}^*(J) - \mu n > n/k$ choices for each. Since $c \gg \mu$, there is some choice of parts $U_i \subseteq V_i$, $2 \leq i \leq k$ of $\mathcal{P}$ such that we obtain at least μn^k cliques intersecting U_i, $i \in [k]$, and at least μn^k cliques intersecting U_1' and U_i, $2 \leq i \leq k$. However, this contradicts the fact that $L_\mathcal{P}^\mu(J_k)$ is transferral-free. Thus Theorem 2.13 implies that J_k has a perfect matching, as required.

2.6. Outline of the proofs

The ideas of our arguments can be roughly organised into the following three groups: regularity, transferrals, applications. Most of the new ideas in this paper pertain to transferrals, but to set the scene we start with regularity. The Szemerédi Regularity Lemma [55] has long been a powerful tool in graph theory. In combination with the blowup lemma of Komlós, Sárközy and Szemerédi [29] it has seen many applications to embeddings of spanning subgraphs (see [32]). Recent developments in hypergraph regularity theory have opened the way towards obtaining analogous results for hypergraphs: the decomposition theory (among other things) was developed independently by Gowers [17] and by Rödl et al. [15, 43, 50, 52], and the blowup lemma by Keevash [22]. Roughly speaking, the decomposition theory allows us to approximate a k-system J on n vertices by a 'reduced' k-system R on m vertices, where m depends on the accuracy of the approximation, but is independent of n. The vertices of R correspond to the parts in some partition of $V(J)$ into 'clusters' of equal size, after moving a small number of vertices to an exceptional set V_0. The edges of R correspond to groups of clusters for which the restriction of the appropriate level of J is well-approximated by a 'dense regular' hypergraph. Furthermore, R inherits from J approximately the same proportional minimum degree sequence. As mentioned earlier, this part of the machinery allows us to reduce the almost perfect matching problem to finding a fractional solution. If J has an extra $o(n)$ in its minimum degree sequence then it is a relatively simple problem in convex geometry to show that R_k has a fractional perfect matching M; moreover, in the absence of a space barrier we can find M even without this extra $o(n)$. Then we partition the clusters of J in proportion to the weights of edges in M, so that each non-zero edge weight in M is associated to a dense regular

k-partite k-graph with parts of equal size (adding a small number of vertices to V_0). It is then straightforward to find almost perfect matchings in each of these k-partite k-graphs, which together constitute an almost perfect matching in J_k.

To find perfect matchings, we start by taking a regularity decomposition and applying the almost perfect matching result in the reduced system. We remove any uncovered clusters, adding their vertices to the exceptional set V_0, so that we have a new reduced system R with a perfect matching M. We also transfer a small number of vertices from each cluster to V_0 so that the edges of M now correspond to dense regular k-partite k-graphs that have perfect matchings (rather than almost perfect matchings). In fact, these k-graphs have the stronger property of 'robust universality': even after deleting a small number of additional vertices, we can embed any bounded degree subhypergraph. (Up to some technicalities not mentioned here, this is the form in which we apply the hypergraph blowup lemma.) Next we greedily find a partial matching that covers V_0, and also removes some vertices from the clusters covered by M, where we take care not to destroy robust universality. Now the edges of M correspond to k-partite k-graphs that are robustly universal, but have slightly differing part sizes. To find a perfect matching, we will remove another partial matching that balances the part sizes and does not destroy robust universality. Then the edges of M will now correspond to k-partite k-graphs that have perfect matchings, and together with the two partial matchings these constitute a perfect matching in J_k.

Transferrals come into play in the step of removing a partial matching to balance the part sizes. Given clusters U and U' and $b \in \mathbb{N}$, a b-fold (U, U')-transferral in (R, M) consists of a pair (T, T') of multisets of edges, with $T \subseteq R$, $T' \subseteq M$ and $|T| = |T'|$, such that every cluster is covered equally often by T and T', except that U is covered b times more in T than in T', and U' is covered b times more in T' than in T. An example is illustrated in Figure 3. Given such a pair (T, T') for which U is too large and U' is too small for our perfect matching strategy, we can reduce the imbalance by b; to achieve this we choose a partial matching in J with edges corresponding to each edge of T (disjoint from all edges chosen so far), and we note for future reference that the perfect matchings corresponding to edges of M chosen in the final step will use one fewer edge corresponding to each edge of T'. We say that (R, M) is (B, C)-irreducible if for any U, U' there is a b-fold (U, U')-transferral of size c for some $b \leq B$ and $c \leq C$ (here we are sticking to the non-partite case for simplicity).

Irreducibility allows us to reduce the cluster imbalances to a constant independent of n, and so find a matching in J_k covering all but a constant number of vertices. Furthermore, it can be expressed in terms of the following geometric condition, alluded to in the introduction. Let

$$X = X(R_k, M) = \{\chi(e) - \chi(e') : e \in R_k, e' \in M\},$$

where $\chi(S)$ denotes the characteristic vector of a set $S \subseteq V(R)$. The required condition is that the convex hull of X should contain a ball of some small constant radius centred at the origin. Thus irreducibility becomes a question of convex geometry. It is also the second point at which space barriers come into play, as our minimum degree assumption implies the existence of such a ball about the origin in the absence of a space barrier. This leads to our first stability result: under the same minimum degree sequence assumption on J needed for a matching in J_k that is almost perfect (i.e. covers all but $o(n)$ vertices), we can in fact find a matching in

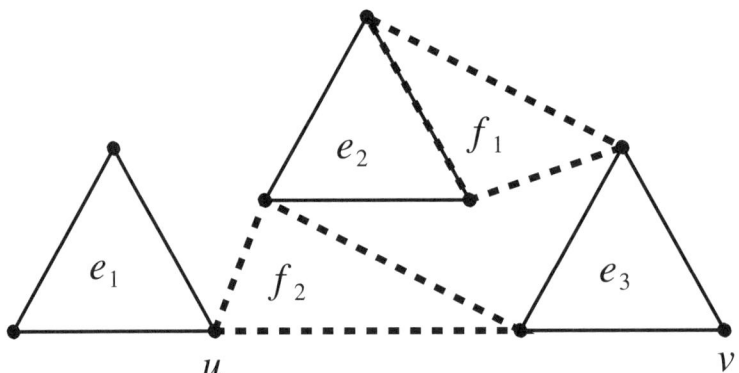

FIGURE 3. An example of a 1-fold (u, v)-transferral in a matched 3-graph (J, M). Here e_1, e_2 and e_3 are edges of the matching M in J, whilst the dashed edges f_1 and f_2 are edges of J alone. Taking $T = \{f_1, f_2\}$ and $T' = \{e_2, e_3\}$, we have $\chi(T) - \chi(T') = \chi(\{u\}) - \chi(\{v\})$.

J_k that covers all but a constant number of vertices, unless J is structurally close to a space barrier construction.

For perfect matchings, we need more precise balancing operations, namely simple transferrals, by which we mean 1-fold transferrals. We can still use general transferrals for most of the balancing operations, but we require simple transferrals to make the cluster sizes precisely equal. We introduce transferral digraphs $D_\ell(R, M)$, where there is an edge from U to U' if and only if (R, M) contains a simple (U, U')-transferral of size at most ℓ. If we now assume for simplicity that J is a k-complex (rather than just a k-system), then it is immediate that every vertex in $D_1(R, M)$ has outdegree at least $\delta^*_{k-1}(R)$. We will prove a structure result for digraphs with linear minimum outdegree, which when combined with irreducibility gives a partition $\mathcal{P}$ of $V(R)$ into a constant number of parts, such that there are simple transferrals between any two vertices in the same part. This is the point at which divisibility barriers come into play. If the robust edge lattice of R_k with respect to $\mathcal{P}$ is incomplete then J_k is structurally close to a divisibility barrier construction. Otherwise, for any pair of parts $P_i, P_j \in \mathcal{P}$ we have some simple (U_i, U_j)-transferrals with $U_i \in P_i$ and $U_j \in P_j$, and the robustness of the edge lattice gives enough of these transferrals to make the cluster sizes precisely equal. Thus we obtain the full stability theorem: under the same minimum degree sequence assumption on J needed for a matching in J_k that is almost perfect, we can find a perfect matching in J_k, unless J is structurally close to a space or divisibility barrier construction.

CHAPTER 3

Geometric Motifs

In this chapter we prove or cite various lemmas that underpin the convex geometry in our theory, and also demonstrate the connection with fractional perfect matchings in hypergraphs. For the classical results cited we refer the reader to the book by Schrijver [53]. We begin with Carathéodory's theorem, in a slightly unusual form. Given points $\mathbf{x}_1, \ldots, \mathbf{x}_s \in \mathbb{R}^d$, we define their *convex hull* as

$$CH(\mathbf{x}_1, \ldots, \mathbf{x}_s) := \left\{ \sum_{j \in [s]} \lambda_j \mathbf{x}_j : \lambda_j \in [0,1], \sum_{j \in [s]} \lambda_j = 1 \right\}.$$

THEOREM 3.1 (Carathéodory's Theorem). *Suppose $X \subseteq \mathbb{R}^d$ and $\mathbf{x} \in CH(X)$. Then there are $\lambda_1, \ldots, \lambda_r \geq 0$ and $\mathbf{x}_1, \ldots, \mathbf{x}_r \in X$ such that*
(a) $\sum_{j \in [r]} \lambda_j = 1$,
(b) *the vectors $\mathbf{x}_j - \mathbf{x}_r$ are linearly independent for each $j \in [r-1]$, and*
(c) $\mathbf{x} = \sum_{j \in [r]} \lambda_j \mathbf{x}_j$.

Note that condition (b) implies that $r \leq d+1$. Indeed, a more standard statement of Theorem 3.1 has the condition $r \leq d+1$ in place of (b); this is commonly proved by first proving our formulation of the theorem. This means that we may write any point in $CH(X)$ as a positive linear combination of a small number of members of X. The following proposition gives conditions under which we can arrange that all of the coefficients λ_j are rationals of small denominator. We say that a real number x is *q-rational* if we can write $x = a/b$ for integers a, b with $1 \leq b \leq q$. We let $\mathbb{Q}_q^d$ denote the set of points $\mathbf{x} \in \mathbb{R}^d$ such that every coordinate of $\mathbf{x}$ is q-rational. Also, given $\mathbf{x} \in \mathbb{R}^d$ and $r \geq 0$ we let

$$B^d(\mathbf{x}, r) = \{\mathbf{z} : \|\mathbf{z} - \mathbf{x}\| \leq r\}$$

denote the ball of radius r centred at $\mathbf{x}$ (we sometimes drop the dimension superscript when this is clear from the context).

PROPOSITION 3.2. *Suppose that $1/q' \ll 1/q, 1/d, 1/k$, and let X be a subset of $\mathbb{Q}_q^d \cap B^d(\mathbf{0}, 2k)$. Then for any $\mathbf{x} \in CH(X) \cap \mathbb{Q}_q^d$ there exist $\lambda_1, \ldots, \lambda_r \geq 0$ and $\mathbf{x}_1, \ldots, \mathbf{x}_r \in X$ such that $r \leq d+1$, each λ_j is q'-rational, $\sum_{j \in [r]} \lambda_j = 1$, and $\mathbf{x} = \sum_{j \in [r]} \lambda_j \mathbf{x}_j$.*

PROOF. By Theorem 3.1 we may choose $z_1, \ldots, z_r \geq 0$ and $\mathbf{x}_1, \ldots, \mathbf{x}_r \in X$ such that $\sum z_j = 1$, the vectors $\mathbf{x}_j - \mathbf{x}_r$ are linearly independent for each $j \in [r-1]$ (and hence $r \leq d+1$) and $\mathbf{x} = \sum_{j \in [r]} z_j \mathbf{x}_j$. We can write $\mathbf{x} = A\mathbf{z} + \mathbf{x}_r$, where A is the d by $r-1$ matrix whose columns are the vectors $\mathbf{x}_i - \mathbf{x}_r$, $i \in [r-1]$, and $\mathbf{z} = (z_1, \ldots, z_{r-1})^\mathsf{T}$. Note that A has rank $r-1$, since its columns are linearly independent. Let $\mathbf{v}_1, \ldots, \mathbf{v}_d$ be the row vectors of A. Then we can choose $S \subseteq [d]$ of size $r-1$ such that the row vectors $\mathbf{v}_j$, $j \in S$ are linearly independent. Let B be the $r-1$ by $r-1$ square matrix with rows $\mathbf{v}_j$ for $j \in S$. Then $\det(B) \neq 0$. Also, since every entry of A has absolute value at most $4k$, we have $\det(B) \leq (r-1)!(4k)^{r-1}$. (A better bound is available from Hadamard's inequality, but it suffices to use this crude bound which follows by estimating each term in the expansion of the determinant.) Now we write $\mathbf{x}' - \mathbf{x}'_r = B\mathbf{z}$, where $\mathbf{x}'$ is the restriction of $\mathbf{x}$ to the coordinates $j \in S$, and $\mathbf{x}'_r$ is defined similarly. Then $\mathbf{z} = B^{-1}(\mathbf{x}' - \mathbf{x}'_r)$, and so every coordinate of $\mathbf{z}$ can be expressed as a fraction with denominator at most $q^2 \det(B) \leq q'$, as required. □

Next we need the classical theorem on the equivalence of vertex and half-plane representations of convex polytopes. This is commonly known as the Weyl-Minkowski theorem; it is also implied by results of Farkas. Given points $\mathbf{v}_1, \ldots, \mathbf{v}_r \in \mathbb{R}^d$, we define their *positive cone* as

$$PC(\{\mathbf{v}_1, \ldots, \mathbf{v}_r\}) := \{\sum_{j \in [r]} \lambda_j \mathbf{v}_j : \lambda_1, \ldots, \lambda_r \geq 0\}.$$

The *Minkowski sum* of two sets $A, B \subseteq \mathbb{R}^d$ is $A + B = \{a + b : a \in A, b \in B\}$.

THEOREM 3.3 (Weyl-Minkowski Theorem). *Let $P \subseteq \mathbb{R}^d$. Then the following statements are equivalent.*
  (i) $P = \{\mathbf{x} \in \mathbb{R}^d : \mathbf{a}_j \cdot \mathbf{x} \geq b_j \text{ for all } j \in [s]\}$ *for some* $\mathbf{a}_1, \ldots, \mathbf{a}_s \in \mathbb{R}^d$ *and* $b_1, \ldots, b_s \in \mathbb{R}$.
  (ii) $P = CH(X) + PC(Y)$ *for some finite sets* $X, Y \subseteq \mathbb{R}^d$.

An important case of Theorem 3.3 is when $P = PC(Y)$ for some finite set $Y \subseteq \mathbb{R}^d$. Then we can write $P = \{\mathbf{x} \in \mathbb{R}^d : \mathbf{a}_j \cdot \mathbf{x} \geq 0 \text{ for all } j \in S\}$ for some $\mathbf{a}_1, \ldots, \mathbf{a}_s \in \mathbb{R}^d$, since we have $\mathbf{0} \in P$, and if $\mathbf{x} \in P$ then $2\mathbf{x} \in P$. The following result of Farkas follows.

LEMMA 3.4 (Farkas' Lemma). *Suppose $\mathbf{v} \in \mathbb{R}^d \setminus PC(Y)$ for some finite set $Y \subseteq \mathbb{R}^d$. Then there is some $\mathbf{a} \in \mathbb{R}^d$ such that $\mathbf{a} \cdot \mathbf{y} \geq 0$ for every $\mathbf{y} \in Y$ and $\mathbf{a} \cdot \mathbf{v} < 0$.*

Our next result exploits the discrete nature of bounded integer polytopes. First we need a convenient description for the faces of a polytope. A *face* of a polytope P is the intersection of P with a set $\mathcal{H}$ of hyperplanes, such that for each $H \in \mathcal{H}$, P is contained in one of the closed halfspaces defined by H. More concretely, consider a finite set $X \subseteq \mathbb{R}^d$ with $\mathbf{0} \in X$. By Theorem 3.3 we can write $CH(X) = \{\mathbf{x} : \mathbf{a}_j \cdot \mathbf{x} \geq b_j \text{ for all } j \in [s]\}$ for some $\mathbf{a}_1, \ldots, \mathbf{a}_s \in \mathbb{R}^d$ and $b_1, \ldots, b_s \in \mathbb{R}$. Let $S = \{j \in [s] : b_j = 0\}$,

$$\Pi_0^X = \{\mathbf{x} : \mathbf{a}_j \cdot \mathbf{x} = 0 \text{ for all } j \in S\} \text{ and } F_0^X = CH(X) \cap \Pi_0^X.$$

Then F_0^X is the (unique) minimum face of $CH(X)$ containing $\mathbf{0}$. Note that in the extreme cases, F_0^X could be all of $CH(X)$, or just the single point $\mathbf{0}$. The following result gives a lower bound for the distance of $\mathbf{0}$ from the boundary of F_0^X.

LEMMA 3.5. *Suppose that $0 < \delta \ll 1/k, 1/d$ and $\mathbf{0} \in X \subseteq \mathbb{Z}^d \cap B^d(\mathbf{0}, 2k)$. Then $\Pi_0^X \cap B^d(\mathbf{0}, \delta) \subseteq CH(X)$*

PROOF. For each X with $\mathbf{0} \in X \subseteq \mathbb{Z}^d \cap B^d(\mathbf{0}, 2k)$ we fix a representation $CH(X) = \{\mathbf{x} : \mathbf{a}_j^X \cdot \mathbf{x} \geq b_j^X \text{ for all } j \in [s^X]\}$ such that $b_j^X \in \mathbb{R}$ and $\mathbf{a}_j^X \in \mathbb{R}^d$ with

$\|\mathbf{a}_j^X\| = 1$ for $j \in [s^X]$. Since $\mathbf{0} \in X$ we must have $b_j^X \leq 0$ for every j and X. Note that there are only finitely many possible choices of X, and each s^X is finite and depends only on X. Choosing δ sufficiently small, we may assume that for any j and X either $b_j^X = 0$ or $b_j^X < -\delta$. Now fix any such set X, and write $F_0^X = CH(X) \cap \Pi_0^X$, where $S^X = \{j \in [s^X] : b_j^X = 0\}$ and $\Pi_0^X = \{\mathbf{x} : \mathbf{a}_j^X \cdot \mathbf{x} = 0 \text{ for all } j \in S^X\}$. Consider any $\mathbf{x} \in \Pi_0^X \setminus CH(X)$. Since $\mathbf{x} \in \Pi_0^X$, $\mathbf{x}$ satisfies all constraints $\mathbf{a}_j^X \cdot \mathbf{x} \geq b_j^X$ for $CH(X)$ with $b_j^X = 0$. Since $\mathbf{x} \notin CH(X)$, $\mathbf{x}$ must fail some constraint $\mathbf{a}_j^X \cdot \mathbf{x} \geq b_j^X$ for $CH(X)$ with $b_j^X \neq 0$. It follows that $\mathbf{a}_j^X \cdot \mathbf{x} < b_j^X < -\delta$. Applying the Cauchy-Schwartz inequality, we have $\delta < |\mathbf{a}_j^X \cdot \mathbf{x}| \leq \|\mathbf{a}_j^X\|\|\mathbf{x}\| = \|\mathbf{x}\|$, as required. □

Now we will demonstrate the connection between convex geometry and fractional perfect matchings, and how these can be obtained under our minimum degree sequence assumption. A *fractional perfect matching* in a hypergraph G is an assignment of non-negative weights to the edges of G such that for any vertex v, the sum of the weights of all edges incident to v is equal to 1. We also require the following notation, which will be used throughout the paper. For any $S \subseteq [n]$, the *characteristic vector* $\chi(S)$ of S is the vector in $\mathbb{R}^n$ given by

$$\chi(S)_i = \begin{cases} 1 & i \in S \\ 0 & i \notin S \end{cases}$$

If G has n vertices, then by identifying $V(G)$ with $[n]$ we may refer to the characteristic vector $\chi(S)$ of any $S \subseteq V(G)$. Whilst $\chi(S)$ is then dependent on the chosen identification of $V(G)$ with $[n]$, the effect of changing this identification is simply to permute the vectors $\chi(S)$ for each $S \subseteq V(G)$ by the same permutation, and all properties we shall consider will be invariant under this isomorphism. So we will often speak of $\chi(S)$ for $S \subseteq V(G)$ without specifying the identification of $V(G)$ and $[n]$. Then a fractional perfect matching is an assignment of weights $w_e \geq 0$ to each $e \in G$ such that $\sum_{e \in G} w_e \chi(e) = \mathbf{1}$; throughout this paper $\mathbf{1}$ denotes a vector of the appropriate dimension in which every coordinate is 1. Thus G has a fractional perfect matching precisely when $\mathbf{1}$ belongs to the positive cone $PC(\chi(e) : e \in G)$. The next lemma shows that this holds under a similar minimum degree sequence assumption to that considered earlier. We include the proof for the purpose of exposition, as a similar argument later in Lemma 7.2 will simultaneously generalise both this statement and a multipartite version of it. The method of proof used in both cases adapts the separating hyperplane argument of Rödl, Ruciński and Szemerédi [49, Proposition 3.1], which was used to prove the existence of a fractional perfect matching in a k-graph in which all but a few sets S of $k-1$ vertices satisfy $d(S) \geq n/k$.

LEMMA 3.6. *Suppose that $k \mid n$, and that J is a k-system on n vertices with*

$$\delta(J) \geq (n, (k-1)n/k, (k-2)n/k, \ldots, n/k).$$

Then J_k admits a fractional perfect matching.

PROOF. Suppose for a contradiction that J_k does not admit a fractional perfect matching. As noted above, this means that $\mathbf{1} \notin PC(\chi(e) : e \in J_k)$. Then by Farkas' Lemma, there is some $\mathbf{a} \in \mathbb{R}^n$ such that $\mathbf{a} \cdot \mathbf{1} < 0$ and $\mathbf{a} \cdot \chi(e) \geq 0$ for every $e \in J_k$. Let $v_1, \ldots, v_n$ be the vertices of J, and let $a_1, \ldots, a_n$ be the corresponding coordinates of $\mathbf{a}$, with the labels chosen so that $a_1 \leq a_2 \leq \cdots \leq a_n$. For any sets S and S' of k

vertices of J, we say that S *dominates* S' if we may write $S = \{v_{i_1}, \ldots, v_{i_k}\}$ and $S' = \{v_{j_1}, \ldots, v_{j_k}\}$ so that $j_\ell \leq i_\ell$ for each $\ell \in [k]$. Note that if S dominates S' then $\mathbf{a} \cdot \chi(S') \leq \mathbf{a} \cdot \chi(S)$; this follows from the fact that the coordinates of $\mathbf{a}$ are increasing.

For each $j \in [n/k]$, let S_j be the set $\{v_j, v_{j+n/k}, v_{j+2n/k}, \ldots, v_{j+(k-1)n/k}\}$. Then the sets S_j partition $V(J)$, so $\sum_{j \in [n/k]} \mathbf{a} \cdot \chi(S_j) = \mathbf{a} \cdot \mathbf{1}$. We claim that there is some edge $e \in J_k$ which is dominated by every S_j. To see this, we let $d_1 = 1$, then apply the minimum degree sequence condition on J to choose $d_2, \ldots, d_k$ greedily so that for each $j \in [k]$ we have $\{v_{d_1}, \ldots v_{d_j}\} \in J$ and $d_j \leq (j-1)n/k + 1$. Then $e := \{v_{d_1}, \ldots, v_{d_k}\}$ is dominated by S_j for each $j \in [n/k]$. We therefore have

$$0 \leq \sum_{j \in [n/k]} \mathbf{a} \cdot \chi(e) \leq \sum_{j \in [n/k]} \mathbf{a} \cdot \chi(S_j) = \mathbf{a} \cdot \mathbf{1} < 0,$$

which is a contradiction. □

CHAPTER 4

Transferrals

To motivate the results of the next two chapters, we start by recalling the proof strategy discussed in Section 2.6. Hypergraph regularity theory (presented in Chapter 6) will enable us to approximate our original k-system by a reduced k-system. To avoid confusion over notation, we should emphasise that all of our transferral results will be applied when J is equal to the reduced system, rather than the original system. We will also have a perfect matching M in J_k, whose edges represent k-tuples of clusters in which we would be able to find a perfect matching, if we were able to make the cluster sizes equal. The role of transferrals is to achieve this by removing a small partial matching. While they are motivated by the proof strategy, the definition and analysis of transferrals does not require any regularity theory (apart from one technical lemma in the next chapter). Suppose that J is a k-graph on $[n]$ and M is a perfect matching in J; for convenience we call the pair (J, M) a *matched k-graph*. Recall that $\chi(e) \in \mathbb{R}^n$ denotes the characteristic vector of $e \subseteq [n]$. If T is a multiset of subsets of $[n]$ we write $\chi(T) = \sum_{e \in T} \chi(e)$, thus identifying T with the multiset in $[n]$ in which the multiplicity of $i \in [n]$ is the number of sets in T containing i, counting with repetition. Also, we say, e.g. 'a multiset T in J' to mean a multiset T of members of J.

4.1. Irreducibility

Now we make some important definitions. Given $b \in \mathbb{N}$, $u, v \in V(J)$ and multisets T in J and T' in M, we say that (T, T') is a *b-fold (u, v)-transferral in (J, M)* if
$$\chi(T) - \chi(T') = b(\chi(\{u\}) - \chi(\{v\})).$$
That is to say that every vertex of J appears equally many times in T as in T', with the exception of u, which appears b times more in T than in T', and v, which appears b times more in T' than in T. An example is shown in Figure 3. Note that if (T, T') is a b-fold (u, v)-transferral in (J, M) then we must have $|T| = |T'|$; we refer to this common size as the *size* of the transferral. We say that (J, M) is *(B, C)-irreducible* if for any $u, v \in V(J)$ there exist $b \leq B$ and $c \leq C$ such that (J, M) contains a b-fold (u, v)-transferral of size c. We also make the following partite version of this definition, in which we only require transferrals within parts. Given a partition $\mathcal{P}$ of $V(J)$ into parts $V_1, \ldots, V_r$, then we say that (J, M) is *(B, C)-irreducible with respect to $\mathcal{P}$* if for any $i \in [r]$ and any $u, v \in V_i$ there exist $b \leq B$ and $c \leq C$ such that (J, M) contains a b-fold (u, v)-transferral of size c. As described in Section 2.6, irreducibility will allow us to reduce the cluster imbalances to a constant, and so find a matching in original system covering all but a constant number of vertices.

It is instructive to consider how irreducibility fails in the space barrier constructions (Construction 2.3, for example as shown in Figure 1). Consider for simplicity the 3-complex $J = J(S, 2)$ when $|S| = 2n/3$ (suppose that $3 \mid n$). By definition we have $|e \cap S| \le 2$ for every $e \in J_3$. So for any perfect matching M in J_3, each $e' \in M$ has $|e' \cap S| = 2$. Thus for any multisets T, T' in J, M with $|T| = |T'|$, the vector $\mathbf{x} = \chi(T) - \chi(T')$ satisfies $\sum_{i \in S} x_i \le 0$. It follows that there are no b-fold (u, v)-transferrals for any $b \in \mathbb{N}$, $u \in S$ and $v \notin S$. Conversely, we will see in the next chapter that if there is no space barrier, then the minimum degree sequence assumption implies irreducibility (with respect to the partition in the partite case). In fact, it implies a geometric condition which is the fundamental property behind irreducibility. To formulate this, we define

$$X = X(J, M) = \{\chi(e) - \chi(e') : e \in J, e' \in M\}.$$

Also, if $\mathcal{P}$ is a partition of V we define

$$\Pi_\mathcal{P} = \{\mathbf{x} \in \mathbb{Z}^n : \mathbf{x} \cdot \chi(U) = 0 \text{ for all } U \in \mathcal{P}\}.$$

The required geometric condition is that the convex hull of X should contain a ball about the origin within $\Pi_\mathcal{P}$. The following lemma shows that this implies irreducibility. (For now we gloss over the fact that B and C depend on n, which is not permissible in our proof strategy; this dependence will be removed by a random reduction in Lemma 5.5.)

LEMMA 4.1. *Suppose that $1/B, 1/C \ll \delta, 1/n, 1/k$. Let V be a set of n vertices and let $\mathcal{P}$ partition V. Suppose that (J, M) is a matched k-graph on V such that $X = X(J, M)$ satisfies $B(\mathbf{0}, \delta) \cap \Pi_\mathcal{P} \subseteq CH(X)$. Then (J, M) is (B, C)-irreducible with respect to $\mathcal{P}$.*

PROOF. Let q be such that $1/B, 1/C \ll 1/q \ll \delta, 1/n, 1/k$. Fix any $U \in \mathcal{P}$ and $u, v \in U$. Let $t = \lceil 2/\delta \rceil$, and let $\mathbf{x} \in \mathbb{Q}^n$ have coordinates $x_u = 1/t$, $x_v = -1/t$ and $x_w = 0$ for $w \ne u, v$. Then $\|\mathbf{x}\| < \delta$ and $\mathbf{x} \in \Pi_\mathcal{P}$, so our hypothesis implies that $\mathbf{x} \in CH(X)$. Applying Proposition 3.2, we may choose $\lambda_1, \dots, \lambda_s \ge 0$ and $\mathbf{x}_1, \dots, \mathbf{x}_s \in X$ such that $s \le n + 1$, each λ_j is q-rational, $\sum_{j \in [s]} \lambda_j = 1$, and $\mathbf{x} = \sum_{j \in [s]} \lambda_j \mathbf{x}_j$. By definition of X, we can choose $e_j \in J$ and $e'_j \in M$ such that $\chi(e_j) - \chi(e'_j) = \mathbf{x}_j$ for $j \in [s]$. Now let the multiset T in J consist of $tq!\lambda_j$ copies of e_j for each $j \in [s]$, and similarly let the multiset T' of edges of M consist of $tq!\lambda_j$ copies of e'_j for each $j \in [s]$. Then

$$\chi(T) - \chi(T') = tq! \sum_{j \in s} \lambda_j \mathbf{x}_j = tq!\mathbf{x}.$$

So the paired multiset (T, T') is a $q!$-fold (u, v)-transferral in (J, M) with size at most $tq! \le C$. Since $U \in \mathcal{P}$ and $u, v \in U$ were arbitrary, we deduce that (J, M) is (B, C)-irreducible with respect to $\mathcal{P}$, as required. □

4.2. Transferral digraphs

To obtain perfect matchings, we need more precise balancing operations, namely *simple transferrals*, by which we mean 1-fold transferrals. We represent these using digraphs (directed graphs), for which we make the following standard definitions. A *digraph* D consists of a vertex set $V(D)$ and an edge set $E(D)$, where the edges are each ordered pairs of vertices; we allow loops (u, u), and 2-cycles $\{(u, v), (v, u)\}$, but do not allow multiple copies of a given ordered pair. We think of an edge $(u, v) \in E(D)$ as being *directed* from u to v; if $(u, v) \in E(D)$ then we say that v is an *outneighbour* of u and that u is an *inneighbour* of v. For any vertex $v \in V(D)$, the *outneighbourhood* $N^+(v)$ is the set of outneighbours of v, and the *inneighbourhood* $N^-(v)$ is the set of inneighbours of v. We also define the *outdegree* $d^+(v) = |N^+(v)|$ and the *indegree* $d^-(v) = |N^-(v)|$, and the *minimum outdegree* $\delta^+(D)$ of D, which is the minimum of $d^+(v)$ over all $v \in D$.

We represent the simple transferrals in (J, M) by the ℓ-*transferral digraph of* (J, M), denoted $D_\ell(J, M)$, where $\ell \in \mathbb{N}$. This is defined to be the digraph on $V(J)$ in which there is an edge from u to v if and only if (J, M) contains a simple (u, v)-transferral of size at most ℓ. For future reference we record here some basic properties of transferral digraphs:

(i) $D_\ell(J, M)$ contains all loops (u, u), as a (u, u)-transferral is trivially achieved by the empty multisets.
(ii) If $\ell \leq \ell'$ then $D_\ell(J, M)$ is a subgraph of $D_{\ell'}(J, M)$.
(iii) If $(u, v) \in D_\ell(J, M)$ and $(v, w) \in D_{\ell'}(J, M)$ then $(v, w) \in D_{\ell+\ell'}(J, M)$. Indeed, if (S, S') is a simple (u, v)-transferral of size at most ℓ and (T, T') is a simple (v, w)-transferral of size at most ℓ', then $(S + T, S' + T')$ is a simple (u, w)-transferral of size at most $\ell + \ell'$.

In the last property we used the notation $A + B$ for the multiset union of multisets A and B. We will also use the notation $\sum_{i \in [t]} A_i$ for the multiset union of multisets A_i, $i \in [t]$, and the notation pA for the multiset union of p copies of A.

Again, it is instructive to consider the examples of the space barrier and divisibility barrier constructions (Constructions 2.3 and 2.8, for example as shown in Figures 1 and 2). As described in the last section, if J is the 3-complex $J(S, 2)$ from Construction 2.3, where $|S| = 2n/3$, and M is a perfect matching in J, then the matched 3-complex (J, M) does not contain a b-fold (u, v)-transferral for any $b \in \mathbb{N}, u \in S$ and $v \notin S$. However, it is not hard to see that (J, M) does contain a simple (u, v)-transferral of size 1 for any other u and v. So for any $\ell \in \mathbb{N}$ the sets S and $V \setminus S$ both induce complete subgraphs in the transferral digraph $D_\ell(J, M)$, and $D_\ell(J, M)$ also contains all edges directed from $V \setminus S$ to S. Now instead consider the k-graph G on vertex set $V = V_1 \cup V_2 \cup V_3$ whose edges are all k-tuples e with $|e \cap V_2| = |e \cap V_3|$ mod 3 (this is example (2) following Construction 2.8), and let M be a perfect matching in G. Then the matched k-graph (G, M) contains a simple (u, v)-transferral of size 1 for any u, v with $u, v \in V_j$ for some $j \in [3]$, but there are no simple (u, v)-transferrals (of any size) with $u \in V_i$ and $v \in V_j$ for $i \neq j$. So for any $\ell \in \mathbb{N}$ the transferral digraph $D_\ell(G, M)$ is the disjoint union of complete digraphs on V_1, V_2 and V_3.

As described in the proof outline earlier, our strategy for finding a perfect matching in a k-complex J uses simple transferrals in the matched k-system (R, M) to 'rebalance' the sizes of clusters of the reduced k-system R. For this we need the transferral digraph $D_\ell(R, M)$ to be complete for some ℓ (or complete on each part of $\mathcal{P}$ if R is $\mathcal{P}$-partite). In this way the transferral digraphs of the two constructions considered above capture the essence of why these constructions do not contain a perfect matching.

Our goal in the remainder of this chapter is to describe the conditions under which, given a matched k-graph (J, M) on V and a partition $\mathcal{P}$ of V, there is some transferral digraph that is complete on each part of $\mathcal{P}$, i.e. there is some ℓ so that $(u, v) \in D_\ell(J, M)$ for every $u, v \in U \in \mathcal{P}$. This is a convenient context that in fact gives more general results than the two cases that are needed for our main theorems, namely (i) when $\mathcal{P}$ consists of a single part V, and (ii) when $\mathcal{P}$ consists of at least

k parts and (J, M) is $\mathcal{P}$-partite. In this section, we prove a structure theorem for general digraphs under a minimum outdegree condition. This introduces a partition $\mathcal{P}'$ that refines $\mathcal{P}$, and we show that irreducibility implies that there is some C' for which $D_{C'}(J, M)$ is complete on the parts of $\mathcal{P}'$. In the final section we show that if there is no divisibility barrier then there is some C for which $D_C(J, M)$ is complete on the parts of $\mathcal{P}$, as required.

We require the following lemma, which states that in a digraph with linear minimum outdegree, one can choose a constant number of vertices so that from any fixed vertex there is a short directed path to one of these chosen vertices.

LEMMA 4.2. *Suppose that* $1/\ell \ll \alpha$, *and let* D *be a digraph on* n *vertices. Then we may choose vertices* $v_1, \ldots, v_t \in V(D)$ *and sets* $V_1, \ldots, V_t \subseteq V(D)$ *such that*
  (a) *for any* j *and any* $u \in V_j$ *there exists a directed path from* u *to* v_j *of length at most* ℓ,
  (b) *the sets* $V_1, \ldots, V_t$ *partition* $V(D)$, *and*
  (c) $|V_j| \geq \delta^+(D) - \alpha n$ *for each* j.

PROOF. Let γ' satisfy $1/(\ell - 1) \ll \gamma' \ll \alpha$. Clearly we may assume that $\delta^+(D) \geq \alpha n$. For any $S \subseteq V(D)$, we let $L(S)$ be the number of edges of D that leave S, i.e. are of the form (u, v) with $u \in S$ and $v \in \overline{S} = V(D) \setminus S$. We require the following claim.

CLAIM 4.3. *Let* $1/(\ell-1) \leq \gamma \leq \gamma'$, *and let* $S \subseteq V(D)$ *be such that* $|S| \geq \alpha n$ *and* $L(S) \leq \gamma n^2$. *Then we may choose* $v \in S$ *and* $S' \subseteq S$ *such that* $|S'| > \delta^+(D) - \alpha n$, $L(S \setminus S') \leq 2\gamma n^2$, *and for any* $u \in S'$ *there is a path from* u *to* v *in* D *of length at most* $\ell - 1$.

To prove the claim, we start by showing that there must be some $v \in S$ with $d_S^-(v) := |N^-(v) \cap S| \geq \delta^+(D) - \alpha n$. To see this, we note that $\sum_{v \in S} d_S^-(v) \geq \sum_{v \in S} d^+(v) - L(S) \geq |S|\delta^+(D) - \gamma n^2$. Thus by averaging we can choose $v \in S$ with $d_S^-(v) \geq \delta^+(D) - \gamma n^2/|S| \geq \delta^+(D) - \alpha n$. Now consider the 'iterated inneighbourhood' N_j^- of v for $j \geq 1$, defined as the set of vertices $u \in S$ such that there exists a path from u to v in D of length at most j. Note that $N_1^- \subseteq N_2^- \subseteq \ldots$, so we can choose some $j \leq 1/\gamma$ for which $|N_{j+1}^-| \leq |N_j^-| + \gamma n$. Now we set $S' = N_j^-$. By construction this satisfies the required property of paths (since $j \leq 1/\gamma \leq \ell - 1$). We also have $|S'| \geq |N_1^-| > \delta^+(D) - \alpha n$, so it remains to estimate $L(S \setminus S')$. Any edge leaving $S \setminus S'$ is either an edge leaving S or an edge from $S \setminus S'$ to S'. By assumption there are at most γn^2 edges of the first type. Also, for any edge (u, v) of the second type we have $u \in N_{j+1}^- \setminus N_j^-$, so there are at most $|N_{j+1}^- \setminus N_j^-|n \leq \gamma n^2$ such edges. This completes the proof of Claim 4.3.

Returning to the proof of the lemma, we apply the claim to $S = V(D)$ with $\gamma = 1/(\ell - 1)$, obtaining $v \in S$ and $S' \subseteq S$ with the stated properties. We set $v_1 = v$ and $V_1 = S'$. Then we repeatedly apply the claim for each $j \geq 1$ to $S = V(D) \setminus (V_1 \cup \cdots \cup V_j)$ with $\gamma = 2^j/(\ell - 1)$; in each application we obtain $v \in S$ and $S' \subseteq S$ for the current set S and we set $v_{j+1} = v$ and $V_{j+1} = S'$. Since $\delta^+(D) \geq \alpha n$, we can repeat this process until we reach some $t \geq 1$ for which $S = V(D) \setminus (V_1 \cup \cdots \cup V_t)$ has $|S| < \delta^+(D)$. Then any $u \in S$ must have an outneighbour $u' \in V(D) \setminus S$, i.e. $u' \in V_j$ for some $j \in [t]$. By choice of V_j there is a path of length at most $\ell - 1$ from u' to v_j, so we have a path of length at most ℓ from u to v_j. For each such u, choose any such j arbitrarily, and add u to V_j. Then the vertices $v_1, \ldots, v_t$ and the new sets $V_1, \ldots, V_t$ satisfy the properties required for the lemma. □

Now we define the structure we will find in our digraphs. Suppose D is a digraph. We say that $S \subseteq V(D)$ is a *dominated set in D* if (u,v) is an edge of D for every $u \in V(D)$ and $v \in S$. Now let $\mathcal{P}$ be a partition of $V(D)$ into parts $V_1, \ldots, V_r$. We say that $\mathcal{S} = (S_1, \ldots, S_r)$ with $S_1 \subseteq V_1, \ldots, S_r \subseteq V_r$ is a *dominated $\mathcal{P}$-tuple* if each S_j is a dominated set in $D[V_j]$. If such an $\mathcal{S}$ exists, we say that $\mathcal{P}$ is a *dominated partition* of D. The *ℓth power* of D is the digraph D^ℓ with vertex set $V(D)$ where $(u,v) \in D^\ell$ if and only if there exists a path from u to v in D of length at most ℓ. Now we can state our structural result, which is that some power of D admits a dominated partition.

COROLLARY 4.4. *Suppose that $1/\ell \ll \alpha$, and let D be a digraph on n vertices. Then D^ℓ admits a dominated partition $\mathcal{P}$ in which each part has size at least $\delta^+(D) - \alpha n$.*

PROOF. Apply Lemma 4.2 to obtain sets $V_1, \ldots, V_t \subseteq V(D)$ and $v_1, \ldots, v_t \in V(D)$ for some t. Let $\mathcal{P}$ be the partition of $V(D)$ into parts $V_1, \ldots, V_t$. So each part of $\mathcal{P}$ has size at least $\delta^+(D) - \alpha n$. Now, for any $j \in [t]$ and any $u \in V_j$ there is a path in D from u to v_j of length at most ℓ, and therefore an edge in D^ℓ from u to v_j. Then $\mathcal{S} := (\{v_1\}, \ldots, \{v_t\})$ is a dominated $\mathcal{P}$-tuple in D^ℓ, as required. □

Now we show that in combination with irreducibility, a dominated partition $\mathcal{P}$ has the property that some transferral digraph is complete on every part of $\mathcal{P}$. We need the following lemma.

LEMMA 4.5. *Let (J, M) be a matched k-graph, $b, c, c' \in \mathbb{N}$, and vertices $u, v \in V(J)$ be such that (J, M) contains a b-fold (u,v)-transferral of size c, and (J, M) contains a simple (v,u)-transferral of size c'. Then (J, M) contains a simple (u,v)-transferral of size $(b-1)c' + c$.*

PROOF. By definition of transferrals, there are multisets T_1, T_2 in J and T_1', T_2' in M such that $|T_1| = |T_1'| = c$, $|T_2| = |T_2'| = c'$,
$$\chi(T_1) - \chi(T_1') = b(\chi(\{u\}) - \chi(\{v\})), \text{ and } \chi(T_2) - \chi(T_2') = \chi(\{v\}) - \chi(\{u\}).$$
Let $T = T_1 + (b-1)T_2$ and $T' = T_1' + (b-1)T_2'$. Then $|T| = |T'| = (b-1)c' + c$, and
$$\chi(T) - \chi(T') = \chi(T_1) - \chi(T_1') + (b-1)(\chi(T_2) - \chi(T_2')) = \chi(\{u\}) - \chi(\{v\}),$$
so (T, T') is the desired transferral. □

Finally, we show the required completeness property of the transferral digraph.

LEMMA 4.6. *Suppose that $1/\ell' \ll 1/B, 1/C, 1/\ell \ll \alpha, \gamma$. Let $\mathcal{P}'$ be a partition of a set V of size n and (J, M) be a matched k-graph on V which is (B, C)-irreducible with respect to $\mathcal{P}'$. Suppose also that $\delta^+(D_\ell(J,M)[U]) \geq \gamma n$ for every $U \in \mathcal{P}'$. Then $D_{\ell^2}(J, M)$ admits a partition $\mathcal{P}$ refining $\mathcal{P}'$ in which every part has size at least $(\gamma - \alpha)n$, and $D_{\ell'}(J,M)[U]$ is complete for every $U \in \mathcal{P}$.*

PROOF. Applying Corollary 4.4 to $D_\ell(J,M)[U]$ for each $U \in \mathcal{P}'$, we see that $D_\ell(J,M)^\ell$ admits a dominated partition $\mathcal{P}$ refining $\mathcal{P}'$ in which every part has size at least $(\gamma - \alpha)n$. We have $D_\ell(J,M)^\ell \subseteq D_{\ell^2}(J,M)$ by property (iii) of transferral digraphs, so $\mathcal{P}$ is also a dominated partition for $D_{\ell^2}(J,M)$. Now suppose $U \in \mathcal{P}$

and $S \subseteq U$ is the dominated set for $D_{\ell^2}(J, M)[U]$. Consider any $u, v \in U$. We need to show that there is a simple (u, v)-transferral of size at most ℓ'. Fix any $w \in S$. By choice of $\mathcal{P}$, we have a simple (u, w)-transferral and a simple (v, w)-transferral each of size at most ℓ^2. By irreducibility, we also have a b-fold (w, v)-transferral of size c for some $b \leq B$ and $c \leq C$. Applying Lemma 4.5, we have a simple (w, v)-transferral of size $(b-1)\ell^2 + c$. Combining this with the simple (u, w)-transferral, we obtain a simple (u, v)-transferral of size at most $b\ell^2 + c \leq \ell'$. □

4.3. Completion of the transferral digraph

In this final section, we show that if in addition to the previous assumptions there is no divisibility barrier, then we can strengthen the previous structure to obtain a transferral digraph that is complete on every part of the original partition. We start with a useful consequence of irreducibility in the following lemma, which allows us to extend a pair of multisets in J and M to a pair of multisets that cover the same vertices (including multiplicities). This corresponds to the geometric intuition that we can use a ball about the origin to counterbalance any given vector. We need to impose the following property on (J, M) that will be automatically satisfied in our applications: we say that (J, M) is $\mathcal{P}$-proper if for any $e \in J$ there is $e' \in M$ with $\mathbf{i}_{\mathcal{P}}(e') = \mathbf{i}_{\mathcal{P}}(e)$.

LEMMA 4.7. *Suppose that $1/C' \ll 1/B, 1/C, 1/k$, let $\mathcal{P}$ be a partition of a set of vertices V, and let (J, M) be a matched k-graph on V which is $\mathcal{P}$-proper and (B, C)-irreducible with respect to $\mathcal{P}$. Then for any multisets A in J and A' in M with $|A|, |A'| \leq C$ there exist multisets T in J and T' in M such that $A \subseteq T$, $A' \subseteq T'$, $|T| = |T'| \leq C'$ and $\chi(T) = \chi(T')$.*

PROOF. First we find multisets S in J and S' in M such that $A \subseteq S$, $A' \subseteq S'$ and $\sum_{e \in S} \mathbf{i}_{\mathcal{P}}(e) = \sum_{e' \in S'} \mathbf{i}_{\mathcal{P}}(e')$. To do this, we let $S = A + A'$ and $S' = A^* + A'$, where A^* is formed by taking some $e' \in M$ with $\mathbf{i}_{\mathcal{P}}(e') = \mathbf{i}_{\mathcal{P}}(e)$ for every $e \in A$ (which exists because (J, M) is $\mathcal{P}$-proper). Since $\sum_{e \in S} \mathbf{i}_{\mathcal{P}}(e) = \sum_{e' \in S'} \mathbf{i}_{\mathcal{P}}(e')$ we may label the vertices of S and S' as $u_1, \ldots, u_{k|S|}$ and $v_1, \ldots, v_{k|S|}$ respectively so that u_i and v_i lie in the same part of $\mathcal{P}$ for each $i \in [k|S|]$. By irreducibility, for each $i \in [k|S|]$ there is a b-fold (u_i, v_i)-transferral in (J, M) of size c for some $b \leq B$ and $c \leq C$. Combining $B!/b$ copies of this transferral, we obtain a $B!$-fold (u_i, v_i)-transferral (T_i, T'_i) of size at most $CB!$. Let $T = B!S + \sum_{i \in [k|S|]} T_i$ and $T' = B!S' + \sum_{i \in [k|S|]} T'_i$. Then $|T| = |T'| \leq CB! + 2Ck \cdot CB! \leq C'$. Also,

$$\chi(T) - \chi(T') = B!(\chi(S) - \chi(S')) + \sum_{i \in [k|S|]} \chi(T_i) - \chi(T'_i)$$

$$= B! \sum_{i \in [k|S|]} \left(\chi(\{u_i\}) - \chi(\{v_i\}) - \chi(\{u_i\}) + \chi(\{v_i\})\right) = 0,$$

which completes the proof. □

Next we need a simple proposition that allows us to efficiently represent vectors in a lattice, in that we have bounds on the number of terms and size of the coefficients in the representation.

PROPOSITION 4.8. *Suppose that $1/s \ll 1/k, 1/d$. Let $X \subseteq \mathbb{Z}^d \cap B^d(\mathbf{0}, 2k)$, and let L_X be the sublattice of $\mathbb{Z}^d$ generated by X. Then for any vector $\mathbf{x} \in L_X \cap$*

$B^d(\mathbf{0}, 2k)$ we may choose multisets S_1 and S_2 of elements of X such that $|S_1|, |S_2| \leq s$ and $\sum_{\mathbf{v} \in S_1} \mathbf{v} - \sum_{\mathbf{v} \in S_2} \mathbf{v} = \mathbf{x}$.

PROOF. The number of pairs $(X, \mathbf{x})$ as in the statement of the proposition depends only on k and d. Furthermore, for any such pair $(X, \mathbf{x})$ we may choose multisets S_1 and S_2 of elements of X such that $\sum_{\mathbf{v} \in S_1} \mathbf{v} - \sum_{\mathbf{v} \in S_2} \mathbf{v} = \mathbf{x}$. Let $s_{X,\mathbf{x}}$ be minimal such that we may do this with $|S_1|, |S_2| \leq s_{X,\mathbf{x}}$. Since $1/s \ll 1/k, 1/d$, we may assume that $s \geq \max_{(X,\mathbf{x})} s_{X,\mathbf{x}}$. Thus for any X and $\mathbf{x}$ we may express $\mathbf{x}$ in the required manner. □

After these preparations, we are ready to prove the main lemma of this chapter. Suppose that $\mathcal{P}'$ is a partition of a set V and $\mathcal{P}$ is a partition into d parts refining $\mathcal{P}'$. Given a k-graph G on V, we define the *edge lattice* $L_{\mathcal{P}}(G) \subseteq \mathbb{Z}^d$ to be the lattice generated by all vectors $\mathbf{i}_{\mathcal{P}}(e)$ with $e \in G$. (Note that this definition is similar to that of the robust edge lattices defined earlier; indeed the purpose of robustness is for the same edge lattice to be inherited by the reduced graph.) Recall that a lattice $L \subseteq \mathbb{Z}^d$ is complete with respect to $\mathcal{P}'$ if $L_{\mathcal{P}\mathcal{P}'} \subseteq L \cap \Pi^d$, where $L_{\mathcal{P}\mathcal{P}'} \subseteq \mathbb{Z}^d \cap \Pi^d$ is the lattice generated by all differences of basis vectors $\mathbf{u}_i - \mathbf{u}_j$ for which V_i, V_j are contained in the same part of $\mathcal{P}'$ and $\Pi^d = \{\mathbf{x} \in \mathbb{R}^d : \sum_{i \in [d]} x_i = 0\}$.

LEMMA 4.9. *Suppose that $1/C' \ll 1/B, 1/C, 1/d, 1/k$. Let $\mathcal{P}'$ be a partition of a set of vertices V, and let (J, M) be a matched k-graph on V which is $\mathcal{P}'$-proper and (B, C)-irreducible with respect to $\mathcal{P}'$. Suppose that $\mathcal{P}$ is a partition of V into d parts $V_1, \ldots, V_d$ which refines $\mathcal{P}'$ such that $D_C(J, M)[V_j]$ is complete for each $j \in [d]$ and $L_{\mathcal{P}}(J)$ is complete with respect to $\mathcal{P}'$. Then (J, M) is $(1, C')$-irreducible with respect to $\mathcal{P}'$, i.e. $D_{C'}(J, M)[U]$ is complete for each $U \in \mathcal{P}'$.*

PROOF. Introduce new constants with $1/C' \ll 1/C_2 \ll 1/C_1 \ll 1/C_0 \ll 1/B, 1/C, 1/d, 1/k$. Fix any $i, j \in [d]$ with $i \neq j$ such that V_i and V_j are both subsets of the same part of $\mathcal{P}'$ (if no such i, j exist then there is nothing to prove). Then $\mathbf{x} := \mathbf{u}_i - \mathbf{u}_j \in L_{\mathcal{P}\mathcal{P}'} \subseteq L_{\mathcal{P}}(J)$ by assumption. By Proposition 4.8 we may therefore choose multisets S and T in J such that $\mathbf{i}_{\mathcal{P}}(S) - \mathbf{i}_{\mathcal{P}}(T) = \mathbf{x}$ and $|S|, |T| \leq C_0$. Note that

$$\sum_{j \in [d]} i_{\mathcal{P}}(S)_j - \sum_{j \in [d]} i_{\mathcal{P}}(T)_j = \sum_{j \in [d]} x_j = 0,$$

$\sum_{j \in [d]} i_{\mathcal{P}}(S)_j = k|S|$ and $\sum_{j \in [d]} i_{\mathcal{P}}(T)_j = k|T|$, so S and T are of equal size. So we may label the vertices of S as $u_1, \ldots, u_r$ and T as $v_1, \ldots, v_r$ so that $u_1 \in V_i$, $v_1 \in V_j$, and u_ℓ and v_ℓ lie in the same part of $\mathcal{P}$ for every $2 \leq \ell \leq r$. Since $D_C(J, M)[V_\ell]$ is complete for each $\ell \in [r]$, for each $2 \leq \ell \leq r$ we may choose (T_ℓ, T'_ℓ) to be a simple (u_ℓ, v_ℓ)-transferral of size at most C. Then

$$\chi(S) - \chi(T) = \sum_{\ell \in [r]} \chi(\{u_\ell\}) - \chi(\{v_\ell\}) = \chi(\{u_1\}) - \chi(\{v_1\}) + \sum_{2 \leq \ell \leq r} \chi(T_\ell) - \chi(T'_\ell).$$

Now, by Lemma 4.7 there exist multisets A_1 in J and A'_1 in M such that $|A_1| = |A'_1| \leq C_1$, $S \subseteq A_1$ and $\chi(A_1) = \chi(A'_1)$. Let $A_2 = A_1 - S + T$ be formed by deleting S from A_1 and replacing it by T. Then $\chi(A_2) - \chi(A'_1) = \chi(A_1) - \chi(S) + \chi(T) - \chi(A'_1) =$

$\chi(T) - \chi(S)$. Finally, let $A_3 = A_2 + \sum_{\ell=2}^{r} T_\ell$ and $A_3' = A_1' + \sum_{\ell=2}^{r} T_\ell'$. Then

$$\chi(A_3) - \chi(A_3') = \chi(A_2) - \chi(A_1') + \sum_{2 \leq \ell \leq r} \chi(T_\ell) - \chi(T_\ell')$$
$$= \chi(\{v_1\}) - \chi(\{u_1\}).$$

So (A_3, A_3') is a simple (v_1, u_1)-transferral of size at most C_2. Since $D_C(J, M)[V_i]$ and $D_C(J, M)[V_j]$ are each complete, we deduce that there exists a simple (v, u)-transferral of size at most $C_2 + 2C \leq C'$ for each $u \in V_i$ and $v \in V_j$. This holds for any i, j such that V_i and V_j are both subsets of the same part U of $\mathcal{P}'$, so $D_{C'}(J, M)[U]$ is complete for each $U \in \mathcal{P}'$. □

Combining Lemmas 4.9 and 4.6 we obtain the following result.

LEMMA 4.10. *Suppose that* $1/\ell' \ll 1/B, 1/C, 1/\ell \ll \alpha, \gamma$. *Let* $\mathcal{P}'$ *be a partition of a set V of size n and (J, M) be a matched k-graph on V that is $\mathcal{P}'$-proper and (B, C)-irreducible with respect to $\mathcal{P}'$. Suppose that* $\delta^+(D_\ell(J, M)[U]) \geq \gamma n$ *for any* $U \in \mathcal{P}'$, *and* $L_\mathcal{P}(J)$ *is complete with respect to $\mathcal{P}'$ for any partition $\mathcal{P}$ of $V(J)$ into d parts of size at least $(\gamma - \alpha)n$ that refines $\mathcal{P}'$. Then $D_{\ell'}(J, M)[U]$ is complete for every* $U \in \mathcal{P}'$.

CHAPTER 5

Transferrals via the minimum degree sequence

In the previous chapter we demonstrated how irreducibility and completeness of the edge lattice imply the existence of all possible simple transferrals. In this chapter we will obtain the same result in the minimum degree sequence settings of our main theorems. In fact we will work in the following setting that simultaneously generalises the partite and non-partite settings.

Let J be a k-system on V and $\mathcal{P}$ be a partition of V into sets $V_1, \ldots, V_r$. An *allocation function* is a function $f : [k] \to [r]$. Intuitively an allocation function should be viewed as a rule for forming an edge of J by choosing vertices one by one; the ith vertex should be chosen from part $V_{f(i)}$ of $\mathcal{P}$. This naturally leads to the notion of the *minimum f-degree sequence*

$$\delta^f(J) := \left(\delta_0^f(J), \ldots, \delta_{k-1}^f(J)\right),$$

where $\delta_j^f(J)$ is the largest m such that for any $\{v_1, \ldots, v_j\} \in J$ with $v_i \in V_{f(i)}$ for $i \in [j]$ there are at least m vertices $v_{j+1} \in V_{f(j+1)}$ such that $\{v_1, \ldots, v_{j+1}\} \in J$. So $\delta_{i-1}^f(J)$ is a lower bound on the number of choices for the ith vertex in the process of forming an edge of J described above. For any allocation function f, the index vector $\mathbf{i}(f)$ of f is the vector

$$\mathbf{i}(f) := (|f^{-1}(1)|, \ldots, |f^{-1}(r)|) \in \mathbb{Z}^r$$

whose jth coordinate is the cardinality of the preimage of j under f. So any edge $e \in J_k$ formed by the described process will have $\mathbf{i}(e) = \mathbf{i}(f)$.

Let I be a multiset of index vectors with respect to $\mathcal{P}$ such that $\sum_{j \in [r]} i_j = k$ for each $\mathbf{i} = (i_1, \ldots, i_r)$ in I. Then we may form a multiset F of allocation functions $f : [k] \to [r]$ as follows. For each $\mathbf{i} \in I$ (with repetition) choose an allocation function f with $\mathbf{i}(f) = \mathbf{i}$, and include in F each of the $k!$ permutations f^σ for $\sigma \in \mathrm{Sym}_k$ (again including repetitions), where $f^\sigma(i) = f(\sigma(i))$ for $i \in [k]$. Note that the multiset F so obtained does not depend on the choices of allocation function f. Also observe that $|F| = k!|I|$, and for any $f \in F$ with $\mathbf{i}(f) = \mathbf{i} \in I$, the multiplicity of f in F is $m_\mathbf{i} \prod_{j \in [r]} i_j!$, where $m_\mathbf{i}$ is the multiplicity of $\mathbf{i}$ in I. If F can be obtained in this manner, then we say that F is an *allocation*, and we write $I(F)$ for the multiset I from which F was obtained. We say that an allocation F is (k,r)-*uniform* if for every $i \in [k]$ and $j \in [r]$ there are $|F|/r$ functions $f \in F$ with $f(i) = j$. We also say that F is *connected* if there is a connected graph G_F on $[r]$ such that for every $jj' \in E(G_F)$ and $i, i' \in [k]$ with $i \neq i'$ there is $f \in F$ with $f(i) = j$ and $f(i') = j'$.

All the allocations F considered in this paper will have the property that $I(F)$ is a set (as opposed to a multiset). In this case we have $I(F) = \{\mathbf{i}(f) : f \in F\}$, and so $|I| \leq r^k$ and $|F| \leq k!|I| \leq k!r^k$. However, we allow the possibility that $I(F)$ is a multiset as this may be useful for future applications. In this case, we shall usually bound $|F|$ by a function $D_F(r,k)$. Also, the reader should take care with any statement regarding (e.g.) $\mathbf{i} \in I(F)$ to interpret the statement with multiplicity. We slightly abuse the notation along these lines, writing (e.g.) "the number of edges of index $\mathbf{i}$ is constant over all $\mathbf{i} \in I(F)$" to mean that the number of edges of index $\mathbf{i}$ is a constant multiple of the multiplicity of $\mathbf{i}$ in $I(F)$.

For an allocation F, a k-system J on V is $\mathcal{P}F$-partite if for any $j \in [k]$ and $e \in J_j$ there is some $f \in F$ so that $e = \{v_1, \ldots, v_j\}$ with $v_i \in V_{f(i)}$ for $i \in [j]$ (so every edge of J can be constructed through the process above for some $f \in F$). The *minimum F-degree sequence* of J is then defined to be

$$\delta^F(J) := (\delta_0^F(J), \ldots, \delta_{k-1}^F(J)),$$

where $\delta_j^F(J) = \min_{f \in F} \delta_j^f(J)$. We note two special cases of this definition that recover our earlier settings. In the case $r = 1$ (the non-partite setting), there is a unique function $f : [k] \to [1]$. Let F consist of $k!$ copies of f; then J is $\mathcal{P}F$-partite and $\delta^F(J) = \delta(J)$ is the (usual) minimum degree sequence. In the case when $r \geq k$ and J is $\mathcal{P}$-partite we can instead choose F to consist of all injective functions from $[k]$ to $[r]$, and then being $\mathcal{P}$-partite is equivalent to being $\mathcal{P}F$-partite and $\delta^F(J) = \delta^*(J)$ is the partite minimum degree sequence. Note that in both cases F is (k,r)-uniform and connected.

Recall that we established a geometric criterion for irreducibility in Lemma 4.1. Our first lemma states that if there is no space barrier then this criterion holds under our generalised minimum degree assumption. Note we have slightly altered the formulation of the space barrier from that used earlier, so that we can state the lemma in the more general context of k-systems; it is not hard to demonstrate that in the case of k-complexes it is an essentially equivalent condition (up to changing the value of β). We say that (J, M) is a *matched k-system* if J is a k-system and M is a perfect matching in J_k. In this case we write $X = X(J,M) = \{\chi(e) - \chi(e') : e \in J_k, e' \in M\}$ and $D_\ell(J, M)$ to mean $D_\ell(J_k, M)$.

LEMMA 5.1. *Suppose that $\delta \ll 1/n \ll \alpha \ll \beta, 1/D_F, 1/r, 1/k$, and $k \mid n$. Let $\mathcal{P}'$ be a partition of a set V into sets $V_1, \ldots, V_r$ each of size n, F be a (k,r)-uniform connected allocation with $|F| \leq D_F$, and (J, M) be a $\mathcal{P}'F$-partite matched k-system on V such that*

$$\delta^F(J) \geq \left(n, \left(\frac{k-1}{k} - \alpha\right)n, \left(\frac{k-2}{k} - \alpha\right)n, \ldots, \left(\frac{1}{k} - \alpha\right)n\right).$$

Suppose that for any $p \in [k-1]$ and sets $S_i \subseteq V_i$ such that $|S_i| = pn/k$ for each $i \in [r]$ we have $|J_{p+1}[S]| \geq \beta n^{p+1}$, where $S = S_1 \cup \cdots \cup S_r$. Then

$$B^{rn}(\mathbf{0}, \delta) \cap \Pi_{\mathcal{P}'} \subseteq CH(X(J,M)).$$

We first sketch the main ideas of the proof for the non-partite case (where $r = 1$, and $\mathcal{P}'$ has only one part, namely V). Suppose for a contradiction that $B^n(\mathbf{0}, \delta) \cap \Pi^k \not\subseteq CH(X)$, where $X = X(J, M)$. We apply Lemma 3.5 to deduce that then there must exist some $\mathbf{a} \in \mathbb{R}^n$ such that $\mathbf{a} \cdot \mathbf{x} \leq 0$ for every $\mathbf{x} \in X$ and $\mathbf{a}$ is not constant on V (i.e. $\mathbf{a}$ is not a constant multiple of $\mathbf{1}$). We use this $\mathbf{a}$ to partition V into k parts $V_1, \ldots, V_k$. Indeed, this partition is chosen so that $|V_1| = n/k - \alpha n$, $|V_2| = |V_3| = \cdots = |V_{k-1}| = n/k$, $|V_k| = n/k + \alpha n$, and moreover for any $i < j$ and any vertices $u \in V_i$ and $v \in V_j$ we have $a_u \leq a_v$, where a_u and a_v are the coordinates of **a** corresponding to u and v respectively. Let u_j be the maximum value of **a** on part V_j.

The reason for choosing parts of these sizes is that, due to our assumption on the minimum degree sequence of J, given any edge e of J_t (for some $t \in [k-1]$) we may greedily extend e to an edge $e \cup \{v_{t+1}, \ldots, v_k\} \in J_k$ in which $v_j \in V_{k-j+1}$ attains at least the maximum value u_{k-j+1} of **a** on V_{k-j+1}. We use this fact repeatedly in the proof.

Let $v_n \in V_k$ be a vertex of V with maximum **a**-coordinate; then greedily extending $\{v_n\} \in J_1$ to an edge $e \in J_k$ as described above gives $\mathbf{a} \cdot \chi(e) \geq \sum_{j \in [k]} u_j =: U$. In particular, the definition of **a** then implies that any edge $e' \in M$ must have $\mathbf{a} \cdot \chi(e') \geq U$. However, since M is a perfect matching in J_k we must have $\sum_{e' \in M} \mathbf{a}' \cdot \chi(e') = \mathbf{a}' \cdot \mathbf{1} \approx U|M'|$. Indeed, the final approximation would be an equality if every V_j had the same size. Since $\mathbf{a}' \cdot \chi(e') \geq \mathbf{a}' \cdot \chi(e)$ for any $e \in J_k$, we conclude that $\mathbf{a}' \cdot \chi(e') \approx \mathbf{a} \cdot \chi(e') \approx U$ for almost every $e' \in M$, and therefore that $\mathbf{a}'$ is close to **a** (this argument is formalised in Claim 5.2).

In particular, we find that $\mathbf{a}'$ is not constant on V. So we may choose some $p \in [k-1]$ so that $u_{p+1} - u_p \gg u_k - u_{p+1}$, that is, the gap between the values of $\mathbf{a}'$ on V_{p+1} and V_p is much greater than the gaps between the values of $\mathbf{a}'$ on $V_{p+1}, \ldots, V_k$. This is the point where we use our assumption that J is far from a space barrier: taking $S = \bigcup_{j > p} V_j$, we can find an edge $e^* \in J_{k-p+1}[S]$. We greedily extend e^* to an edge $e \in J_k$ as described earlier, giving $\mathbf{a} \cdot \chi(e) \geq (k-p+1)u_{p+1} + \sum_{j \in [p-1]} u_j$, which by choice of p is significantly greater than U. For any $e' \in M$ with $\mathbf{a} \cdot \chi(e') \approx U$ this gives $\mathbf{a} \cdot \chi(e) > \mathbf{a} \cdot \chi(e')$, contradicting the choice of **a** for $\mathbf{x} = \chi(e) - \chi(e') \in X$.

We now give the full details of the proof.

Proof of Lemma 5.1. We start by applying Lemma 3.5 to $X = X(J,M) \subseteq \mathbb{Z}^{rn} \cap B^{rn}(\mathbf{0}, 2k)$. This gives $\Pi_0^X \cap B^{rn}(\mathbf{0}, \delta) \subseteq CH(X)$, where $F_0^X = CH(X) \cap \Pi_0^X$ is the minimum face of $CH(X)$ containing **0**. We can write $\Pi_0^X = \{\mathbf{x} \in \mathbb{Z}^{rn} : \mathbf{a} \cdot \mathbf{x} = 0 \text{ for all } \mathbf{a} \in A\}$ for some $A \subseteq \mathbb{R}^{rn}$ such that $\mathbf{a} \cdot \mathbf{x} \leq 0$ for every $\mathbf{x} \in X$ and $\mathbf{a} \in A$. To prove that $B^{rn}(\mathbf{0}, \delta) \cap \Pi_{\mathcal{P}'} \subseteq CH(X)$, it suffices to show that A is contained in the subspace generated by $\{\chi(V_i) : i \in [r]\}$. Suppose for a contradiction that some $\mathbf{a} \in A$ is not in $\langle \{\chi(V_i) : i \in [r]\} \rangle$. Then there is some $i' \in [r]$ such that **a** is not constant on $V_{i'}$.

We label V so that for each $i \in [r]$, the vertices of V_i are labelled $v_{i,1}, \ldots, v_{i,n}$, the corresponding coordinates of **a** are $a_{i,1}, \ldots, a_{i,n}$, and

$$a_{i,1} \leq a_{i,2} \leq \cdots \leq a_{i,n}.$$

It is convenient to assume that we have a strict inequality $\delta_i^F(J) > \left(\frac{k-i}{k} - \alpha\right)n$ for $i \in [k-1]$. This can be achieved by replacing α with a slightly smaller value; we can also assume that $\alpha n \in \mathbb{N}$. Next we partition each V_i into k parts $V_{i,1}, \ldots, V_{i,k}$ as follows. For $i \in [r]$ we let

$$V_{i,j} = \begin{cases} \{v_{i,\ell} : 0 < \ell \leq n/k - \alpha n\} & \text{if } j = 1, \\ \{v_{i,\ell} : (j-1)n/k - \alpha n < \ell \leq jn/k - \alpha n\} & \text{if } 2 \leq j \leq k-1, \\ \{v_{i,\ell} : (k-1)n/k - \alpha n < \ell \leq n\} & \text{if } j = k. \end{cases}$$

Then for each $i \in [r]$ we have

$$|V_{i,j}| = \begin{cases} \frac{n}{k} - \alpha n & \text{if } j = 1, \\ \frac{n}{k} & \text{if } 2 \leq j \leq k-1, \\ \frac{n}{k} + \alpha n & \text{if } j = k. \end{cases} \tag{1}$$

We use these partitions to define a simpler vector $\mathbf{a}'$ that will be a useful approximation to $\mathbf{a}$. For $i \in [r]$ and $j \in [n]$ we let $p(j) \in [k]$ be such that $v_{i,j} \in V_{i,p(j)}$. For $i \in [r]$ and $\ell \in [k]$ we let $u_{i,\ell} = \max\{a_{i,j} : p(j) = \ell\}$. Then we define $\mathbf{a}' \in \mathbb{Z}^{rn}$ by

$$\mathbf{a}' := (a'_{i,j} : i \in [r], j \in [n]), \text{ where } a'_{i,j} = u_{i,p(j)}.$$

Note that $a'_{i,j} \geq a_{i,j}$ for every $i \in [r]$ and $j \in [n]$. We can assume without loss of generality that $u_{i,k} - u_{i,1}$ is maximised over $i \in [r]$ when $i = 1$; then we define

$$d = u_{1,k} - u_{1,1}, \quad U = \frac{1}{r} \sum_{i \in [r], j \in [k]} u_{i,j}, \text{ and } U_f = \sum_{j \in [k]} u_{f(j),j} \text{ for } f \in F.$$

Note that uniformity of F implies that U is the average of U_f over $f \in F$. We will prove the following claim.

CLAIM 5.2. *We have the following properties.*
  (i) *For any $e' \in M$ and $f \in F$ we have $\mathbf{a}' \cdot \chi(e') \geq \mathbf{a} \cdot \chi(e') \geq U_f$ and $\mathbf{a} \cdot \chi(e') \geq U$,*
  (ii) *$d > 0$ and $\sum_{e' \in M}(\mathbf{a}' \cdot \chi(e') - U) \leq rd\alpha n$,*
  (iii) *There is some $e' \in M$ such that $\mathbf{a} \cdot \chi(e') \leq U_f + D_F k d\alpha$ for every $f \in F$,*
  (iv) *For any $i, i' \in [r]$ and $j, j' \in [k]$ we have $|(u_{i,j} - u_{i,j'}) - (u_{i',j} - u_{i',j'})| \leq D_F rk d\alpha$.*

To see property (i), note that the first inequality follows from $a'_{i,j} \geq a_{i,j}$. For the second, we use the minimum f-degree sequence of J to greedily form an edge $e = \{v_{f(k),\ell_k}, \ldots, v_{f(1),\ell_1}\} \in J_k$, starting with $\ell_k = n$, then choosing $\ell_j > jn/k - \alpha n$ for each $j = k-1, k-2, \ldots, 1$. Then $p(\ell_j) \geq j+1$ for $j \in [k-1]$, so $\mathbf{a} \cdot \chi(e) \geq U_f$. Now the inequality follows from the definition of A, which gives $\mathbf{a} \cdot (\chi(e') - \chi(e)) \geq 0$. We noted above that U can be expressed as an average of some U_f's, so $\mathbf{a} \cdot \chi(e') \geq U$. This proves property (i).

To see that $d > 0$, suppose for a contradiction that $u_{i,k} = u_{i,1}$ for $i \in [r]$. Recall that there is some $i' \in [r]$ such that $\mathbf{a}$ is not constant on $V_{i'}$. Let $e' \in M$ be the edge of M containing $v_{i',1}$. Since (J, M) is $\mathcal{P}'F$-partite we can write $e' = \{v_{f(i),n_i}\}_{i \in [k]}$ for some $f \in F$ and $n_i \in [n]$ for $i \in [k]$, where without loss of generality $v_{f(1),n_1} = v_{i',1}$. Then

$$\mathbf{a} \cdot \chi(e') \leq a_{i',1} + \sum_{i=2}^{k} a_{f(i),n_i} < \sum_{i=1}^{k} a_{f(i),n} = U_f.$$

However, this contradicts property (i), so $d > 0$. Next note that $\sum_{e' \in M} \chi(e') = \mathbf{1}$ and $|M| = rn/k$, so by (1) and the definition of d we have

$$\sum_{e' \in M}(\mathbf{a}' \cdot \chi(e') - U) = \mathbf{a}' \cdot \mathbf{1} - |M|U = \sum_{i \in [r], j \in [k]}(|V_i| - n/k)u_{i,j}$$

$$= \sum_{i \in [r]}(u_{i,k} - u_{i,1})\alpha n \leq rd\alpha n,$$

so we have proved property (ii).

Now we may choose some $e' \in M$ with $0 \leq \mathbf{a}' \cdot \chi(e') - U \leq rd\alpha n/|M| = kd\alpha$. Since $U = |F|^{-1} \sum_{f \in F} U_f$, $\mathbf{a}' \cdot \chi(e') \geq \mathbf{a} \cdot \chi(e') \geq U_f$ for any $f \in F$ and $|F| \leq D_F$, we have $0 \leq \mathbf{a}' \cdot \chi(e') - U_f \leq D_F k d\alpha$; this implies (iii), using $\mathbf{a}' \cdot \chi(e') \geq \mathbf{a} \cdot \chi(e')$. It also implies $|U_f - U_{f'}| \leq D_F k d\alpha$ for any $f, f' \in F$. Now, by definition of G_F and invariance of F under Sym_k, for any $i, i' \in [r]$ and $j, j' \in [k]$ such that $ii' \in G_F$, we can choose $f \in F$ with $f(j) = i$ and $f(j') = i'$, and let $f' \in F$ be obtained from f by transposing the values on j and j'. Then $|U_f - U_{f'}| = |(u_{i,j} - u_{i,j'}) - (u_{i',j} - u_{i',j'})| \leq D_F k d\alpha$. Since G_F is connected we obtain statement (iv), so we have proved Claim 5.2.

To continue the proof of Lemma 5.1, we say that an edge $e' \in M$ is *good* if $\mathbf{a}' \cdot \chi(e') \leq U + d\sqrt{\alpha}$. We say that a vertex $v \in V(J)$ is *good* if it lies in a good edge of M. Note that if $v_{i,j}$ is a good vertex and $v_{i,j} \in e' \in M$, then Claim 5.2 (i) gives $0 \leq (\mathbf{a}' - \mathbf{a}) \cdot \chi(e') \leq d\sqrt{\alpha}$, so $a'_{i,j} - a_{i,j} < d\sqrt{\alpha}$. Writing B for the set of bad edges of M, by Claim 5.2 (i) and (ii) we have

$$|B| d\sqrt{\alpha} < \sum_{e' \in B} (\mathbf{a}' \cdot \chi(e') - U) \leq \sum_{e' \in M} (\mathbf{a}' \cdot \chi(e') - U) \leq rd\alpha n,$$

so $|B| < r\sqrt{\alpha} n$. Thus all but at most $rk\sqrt{\alpha} n$ vertices $v \in V(J)$ are good. Next we need another claim.

CLAIM 5.3. *There is some $p \in [k-1]$ and an edge $e^* \in J_{k-p+1}$ in which all vertices are good, such that $\mathbf{a}' \cdot \chi(e^*) \geq (k+1)d\sqrt{\alpha} + \sum_{p \leq j \leq k} u_{f(j),j}$, where $f \in F$ is such that $e^* = \{v_p, \ldots, v_k\}$ with $v_j \in V_{f(j)}$ for $p \leq j \leq k$.*

To prove this claim, we start by choosing p so that the gap $u_{1,p+1} - u_{1,p}$ is considerably larger than $u_{1,k} - u_{1,p+1}$. More precisely, we require

$$u_{1,p+1} - u_{1,p} > k(u_{1,k} - u_{1,p+1}) + (k+2)d\sqrt{\alpha}.$$

Suppose for a contradiction that this is not possible. Then for every $p \in [k-1]$ we have

$$u_{1,k} - u_{1,p} = (u_{1,k} - u_{1,p+1}) + (u_{1,p+1} - u_{1,p}) \leq (k+2)(u_{1,k} - u_{1,p+1} + d\sqrt{\alpha}).$$

Iterating this inequality starting from $u_{1,k} - u_{1,k-1} \leq (k+2)d\sqrt{\alpha}$ we obtain the bound $u_{1,k} - u_{1,p} \leq (k+3)^{k-p} d\sqrt{\alpha}$. However, for $p = 1$ we obtain $d = u_{1,k} - u_{1,1} \leq (k+3)^{k-1} d\sqrt{\alpha}$, which is a contradiction for $\alpha \ll 1/k$. Thus we can choose p as required.

Now consider $S = \bigcup_{i \in [r], p < j \leq k} V_{i,j}$. Note that $|S \cap V_i| = (k-p)n/k + \alpha n$ for each $i \in [r]$. Thus we can apply the assumption that there is no space barrier, which gives at least βn^{k-p+1} edges in $J_{k-p+1}[S]$. At most $rk\sqrt{\alpha} n^{k-p+1}$ of these edges contain a vertex which is not good, so we may choose an edge $e^* \in J_{k-p+1}$ whose vertices are all good and lie in S. Since J is $\mathcal{P}'F$-partite, we can choose $f \in F$ so that $e^* = \{v_{f(p),\ell_p}, \ldots, v_{f(k),\ell_k}\}$ for some $\ell_p, \ldots, \ell_k$, and since $e^* \subseteq S$ we have $\mathbf{a}' \cdot \chi(e^*) \geq \sum_{p \leq j \leq k} u_{f(j),p+1}$. Then by Claim 5.2 (iv) we can estimate

$$\mathbf{a}' \cdot \chi(e^*) - \sum_{p \leq j \leq k} u_{f(j),j} \geq \sum_{p \leq j \leq k} (u_{1,p+1} - u_{1,j}) - D_F r k^2 d\alpha.$$

By choice of p we have

$$\sum_{p \leq j \leq k} (u_{1,p+1} - u_{1,j}) \geq (u_{1,p+1} - u_{1,p}) - (k-p)(u_{1,k} - u_{1,p+1}) > (k+2)d\sqrt{\alpha}.$$

The required bound on $\mathbf{a}' \cdot \chi(e^*)$ follows, so this proves Claim 5.3.

To finish the proof of Lemma 5.1, using the minimum degree sequence of J as above, we greedily extend e^* to an edge $e = \{v_{f(k),\ell_k}, \ldots, v_{f(1),\ell_1}\} \in J_k$ with $\ell_j > jn/k - \alpha n$ for $j = p-1, p-2, \ldots, 1$. This gives $\mathbf{a} \cdot \chi(e \setminus e^*) \geq \sum_{j \in [p-1]} u_{f(j),j}$. Also, since every vertex of e^* is good we have $(\mathbf{a}' - \mathbf{a}) \cdot \chi(e^*) \leq kd\sqrt{\alpha}$. It follows that

$$\mathbf{a} \cdot \chi(e) = \mathbf{a}' \cdot \chi(e^*) - (\mathbf{a}' - \mathbf{a}) \cdot \chi(e^*) + \mathbf{a} \cdot \chi(e \setminus e^*)$$
$$\geq (k+1)d\sqrt{\alpha} + \sum_{p \leq j \leq k} u_{f(j),j} - kd\sqrt{\alpha} + \sum_{j \in [p-1]} u_{f(j),j} = U_f + d\sqrt{\alpha}.$$

On the other hand, for any $e' \in M$ we have $\mathbf{a} \cdot \chi(e) \leq \mathbf{a} \cdot \chi(e')$ by definition of A, and by Claim 5.2 (iii) we can choose e' so that $\mathbf{a} \cdot \chi(e') \leq U_f + D_F kd\alpha$. Thus we have a contradiction to our original assumption that $A \not\subseteq \langle \{\chi(V_i) : i \in [r]\}\rangle$, which proves the Lemma.

Next we need to address a technical complication alluded to earlier, which is that we cannot satisfy the assumption $\delta \ll 1/n$ in Lemma 5.1 by working directly with J. Thus we need the next lemma, which will be proved by taking a random selection of edges in M, that allows us to reduce to a small matched subsystem (J', M') with similar properties to (J, M). For this we need the following definitions.

DEFINITION 5.4. *Suppose that $\mathcal{P}$ partitions a vertex set V into parts $V_1, \ldots, V_r$, that (J, M) is a matched k-graph or k-system on V, and that F is a (k, r)-uniform allocation. We say that M α-represents F if for any $\mathbf{i}, \mathbf{i}'$ in $I(F)$, letting N, N' denote the number of edges $e' \in M$ with index $\mathbf{i}, \mathbf{i}'$ respectively, we have $N' \geq (1 - \alpha)N$. That is, each $\mathbf{i} \in I(F)$ is represented by approximately the same number of edges of M.*

In particular, if M 0-represents F then the number of edges $e' \in M$ with $\mathbf{i}(e') = \mathbf{i}$ is the same for all $\mathbf{i} \in I(F)$; in this case we say that M is F-balanced.

Note that when (J, M) is $\mathcal{P}F$-partite, the fact that M α-represents F implies that (J, M) is $\mathcal{P}$-proper (recall that this was a condition needed for Lemma 4.9).

LEMMA 5.5. *Suppose that $1/n \ll 1/n' \ll \alpha' \ll \alpha \ll \beta' \ll \beta \ll 1/D_F, 1/r, 1/k$. Let $\mathcal{P}$ partition a set V into r parts $V_1, \ldots, V_r$ each of size n. Suppose F is a (k, r)-uniform allocation with $|F| \leq D_F$ and (J, M) is a $\mathcal{P}F$-partite matched k-system on V such that M α'-represents F. Suppose also that*
  (i) $\delta^F(J) \succeq \left(n, \left(\frac{k-1}{k} - \alpha\right)n, \left(\frac{k-2}{k} - \alpha\right)n, \ldots, \left(\frac{1}{k} - \alpha\right)n\right)$, *and*
  (ii) *for any $p \in [k-1]$ and sets $S_i \subseteq V_i$ such that $|S_i| = \lfloor pn/k \rfloor$ for each $i \in [r]$ there are at least βn^{p+1} edges in $J_{p+1}[S]$, where $S := \bigcup_{i \in [r]} S_i$.*

Then for any $a \in [r]$ and $u, v \in V_a$, there exists a set $M' \subseteq M$ such that, writing $V' = \bigcup_{e \in M'} e$, $J' = J[V']$, and $\mathcal{P}'$ for the partition of V' into $V'_i := V_i \cap V'$, $i \in [r]$, we have $u, v \in V'_a$, $|V'_1| = \cdots = |V'_r| = n'$, and (J', M') is a $\mathcal{P}'F$-partite matched k-system on V' such that
  (i) $\delta^F(J') \succeq \left(n', \left(\frac{k-1}{k} - 2\alpha\right)n', \left(\frac{k-2}{k} - 2\alpha\right)n', \ldots, \left(\frac{1}{k} - 2\alpha\right)n'\right)$, *and*
  (ii) *for any $p \in [k-1]$ and sets $S'_i \subseteq V'_i$ such that $|S'_i| = \lfloor pn'/k \rfloor$ for each $i \in [r]$ there are at least $\beta'(n')^{p+1}$ edges in $J'_{p+1}[S']$, where $S' := \bigcup_{i \in [r]} S'_i$.*

The proof of this lemma requires some regularity theory, so we postpone it to the next chapter. Now we can combine Lemma 5.1 and Lemma 5.5, to obtain the following lemma, which shows that our assumptions guarantee irreducibility with the correct dependence of parameters. Note that the condition on M in the final statement is automatically satisfied if J is a complex.

LEMMA 5.6. *Suppose that $1/n \ll 1/\ell \ll 1/B, 1/C \ll \alpha' \ll \alpha \ll \beta \ll 1/D_F, 1/r, 1/k$. Let V be a set of rn vertices, and let $\mathcal{P}'$ partition V into r parts $V_1, \ldots, V_r$ each of size n. Suppose F is a (k,r)-uniform connected allocation with $|F| \leq D_F$, and (J, M) is a $\mathcal{P}'F$-partite matched k-system on V such that M α'-represents F. Suppose also that*

(i) $\delta^F(J) \geq \left(n, \left(\frac{k-1}{k} - \alpha\right)n, \left(\frac{k-2}{k} - \alpha\right)n, \ldots, \left(\frac{1}{k} - \alpha\right)n\right)$, *and*
(ii) *for any $p \in [k-1]$ and sets $S_i \subseteq V_i$ such that $|S_i| = \lfloor pn/k \rfloor$ for each $i \in [r]$ there are at least βn^{p+1} edges in $J_{p+1}[S]$, where $S := \bigcup_{i \in [r]} S_i$.*

Then (J_k, M) is (B, C)-irreducible with respect to $\mathcal{P}'$.

If, in addition, for any $v \in e' \in M$ we have $e' \setminus \{v\} \in J$, then there is a partition $\mathcal{P}$ refining $\mathcal{P}'$ such that $|U| \geq n/k - 2\alpha n$ and $D_\ell(J, M)[U]$ is complete for every $U \in \mathcal{P}$.

PROOF. Choose n', δ with $1/B, 1/C \ll \delta \ll 1/n' \ll \alpha'$ and $kr \mid n'$, fix any $i \in [r]$, $u, v \in V_i$, and let M', V', J' be given by applying Lemma 5.5. Let $\mathcal{P}''$ be the partition of V' into $V' \cap V_1, \ldots, V' \cap V_r$ and $X' = X(J', M') \subseteq \mathbb{R}^{n'}$. Applying Lemma 5.1, we have $B^{rn'}(\mathbf{0}, \delta) \cap \Pi_{\mathcal{P}''} \subseteq CH(X')$. Then by Lemma 4.1 we deduce that (J'_k, M') is (B, C)-irreducible with respect to $\mathcal{P}''$. Since $i \in [r]$ and $u, v \in V_i$ were arbitrary, it follows that (J_k, M) is (B, C)-irreducible with respect to $\mathcal{P}'$. For the final statement we claim that $\delta^+(D_1(J_k, M)[V_q]) \geq \delta^F_{k-1}(J) \geq n/k - \alpha n$ for each $q \in [r]$. Indeed, consider any $v \in V_q$, and let e' be the edge of M containing v. Then we can write $e' = \{v_1, \ldots, v_k\}$ where $v_k = v$ and $v_i \in V_{f(i)}$ for $i \in [k]$ for some $f \in F$. By assumption $e' \setminus \{v\} \in J$, so there are at least $\delta^f_{k-1}(J)$ vertices $u \in V_q$ such that $\{u\} \cup e' \setminus \{v\} \in J$. For each such u, $(\{\{u\} \cup e' \setminus \{v\}\}, \{e'\})$ is a simple (u, v)-transferral in (J_k, M) of size one, so this proves the claim. The conclusion then follows from Lemma 4.6. □

Now we can formulate our main transferral lemma, which is that the minimum degree sequence assumption, with no divisibility or space barrier, implies the existence of all required simple transferrals. The proof is immediate from our previous lemmas.

LEMMA 5.7. *Suppose that $1/n \ll 1/\ell \ll \alpha' \ll \alpha \ll \beta \ll 1/D_F, 1/r, 1/k$. Let V be a set of rn vertices, and let $\mathcal{P}'$ be a partition of V into parts $V_1, \ldots, V_r$ each of size n. Suppose F is a (k,r)-uniform connected allocation with $|F| \leq D_F$, (J, M) is a $\mathcal{P}'F$-partite matched k-system on V such that M α'-represents F, and for any $v \in e' \in M$ we have $e' \setminus \{v\} \in J$. Suppose also that*

(i) $\delta^F(J) \geq \left(n, \left(\frac{k-1}{k} - \alpha\right)n, \left(\frac{k-2}{k} - \alpha\right)n, \ldots, \left(\frac{1}{k} - \alpha\right)n\right)$,
(ii) *for any $p \in [k-1]$ and sets $S_i \subseteq V_i$ such that $|S_i| = \lfloor pn/k \rfloor$ for each $i \in [r]$ there are at least βn^{p+1} edges in $J_{p+1}[S]$, where $S := \bigcup_{i \in [r]} S_i$, and*
(iii) $L_\mathcal{P}(J_k)$ *is complete with respect to $\mathcal{P}'$ for any partition $\mathcal{P}$ of $V(J)$ which refines $\mathcal{P}'$ and whose parts each have size at least $n/k - 2\alpha n$.*

Then $D_\ell(J_k, M)[V_j]$ is complete for each $j \in [r]$. That is, for any $j \in [k]$ and $u, v \in V_j$ the matched k-graph (J_k, M) contains a simple (u,v)-transferral of size at most ℓ.

PROOF. Let ℓ', B and C satisfy $1/\ell \ll 1/\ell' \ll 1/C, 1/B \ll \alpha$. By Lemma 5.6, (J_k, M) is (B, C)-irreducible with respect to $\mathcal{P}'$, and there is a partition $\mathcal{P}$ refining $\mathcal{P}'$ such that $|U| \geq n/k - 2\alpha n$ and $D_{\ell'}(J, M)[U]$ is complete for every $U \in \mathcal{P}$. Then $L_\mathcal{P}(J_k)$ is complete with respect to $\mathcal{P}'$ by assumption, so Lemma 4.9 applies to give the required result. □

CHAPTER 6

Hypergraph Regularity Theory

This chapter contains the technical background that we need when applying the method of hypergraph regularity. Most of the machinery will be quoted from previous work, although we also give some definitions and short lemmas that are adapted to our applications. We start in the first section by introducing our notation, defining hypergraph regularity, and stating the 'regular restriction lemma'. The second section states the Regular Approximation Lemma of Rödl and Schacht. In the third section we state the Hypergraph Blow-up Lemma, and prove a short accompanying result that is adapted to finding perfect matchings. In the fourth section we define reduced k-systems, and develop their theory, including the properties that they inherit degree sequences, and that their edges represent k-graphs to which the Hypergraph Blow-up Lemma applies. The final section contains the proof of Lemma 5.5, using the simpler theory of 'weak regularity' (as opposed to the 'strong regularity' in the rest of the chapter).

6.1. Hypergraph regularity

Let X be a set of vertices, and let $\mathcal{Q}$ be a partition of X into r parts $X_1, \ldots, X_r$. The *index* $i_{\mathcal{Q}}(S)$ of a set $S \subseteq X$ is the multiset in $[r]$ where the multiplicity of $j \in [r]$ is $\mathbf{i}_{\mathcal{Q}}(S)_j = |S \cap X_j|$; we generally write $i(S) = i_{\mathcal{Q}}(S)$ when $\mathcal{Q}$ is clear from the context. Recall that a set $S \subseteq X$ is $\mathcal{Q}$-partite if it has at most one vertex in each part of $\mathcal{Q}$; for such sets $i(S)$ is a set. For any $A \subseteq [r]$ we write X_A for the set $\bigcup_{i \in A} X_i$, and $K_A(X)$ for the complete $\mathcal{Q}$-partite $|A|$-graph on X_A, whose edges are all $\mathcal{Q}$-partite sets $S \subseteq X$ with $i(S) = A$. For a $\mathcal{Q}$-partite set $S \subseteq X$ and $A \subseteq i(S)$, we write $S_A = S \cap X_A$. If H is a $\mathcal{Q}$-partite k-graph or k-system on X, then for any $A \subseteq [r]$ we define $H_A := H \cap K_A(X)$. Equivalently, H_A consists of all edges of H with index A, which we regard as an $|A|$-graph on vertex set X_A. If additionally H is a k-complex, then we write $H_{A \leq}$ for the k-complex on X_A with edge set $\bigcup_{B \subseteq A} H_B$, and $H_{A<}$ for the k-complex on X_A with edge set $\bigcup_{B \subset A} H_B$. Similarly, we write $H_\mathbf{i}$ for the set of edges in H with index vector $\mathbf{i}$. This is a $|\mathbf{i}|$-graph, where $|\mathbf{i}| := \sum_{j \in [r]} i_j$.

To understand the definition of hypergraph regularity it is helpful to start with the case of graphs. If G is a bipartite graph on vertex classes U and V, we say that G is ε-*regular* if for any $U' \subseteq U$ and $V' \subseteq V$ with $|U'| > \varepsilon |U|$ and $|V'| > \varepsilon |V|$ we have $d(G[U' \cup V']) = d(G) \pm \varepsilon$. Similarly, for a k-complex G, an informal statement of regularity is that the restriction of G to any large subcomplex of the 'lower levels' of G has similar densities to G. Formally, let $\mathcal{Q}$ partition a set X into parts $X_1, \ldots, X_r$, and G be an $\mathcal{Q}$-partite k-complex on X. We denote by G_A^* the $|A|$-graph on X_A whose edges are all those $S \in K_A(X)$ such that $S' \in G$ for every strict subset $S' \subset S$. So G_A^* consists of all sets which could be edges of G_A, in the sense that they are supported by edges at 'lower levels'. The *relative density of* G

at A is
$$d_A(G) := \frac{|G_A|}{|G_A^*|};$$
this is the proportion of 'possible edges' (given $G_{A<}$) that are in fact edges of G_A. (It should not be confused with the *absolute density of G at A*, which is $d(G_A) := |G_A|/|K_A(X)|$.)

For any $A \in \binom{[r]}{\leq k}$, we say that G_A is ε-*regular* if for any subcomplex $H \subseteq G_{A<}$ with $|H_A^*| \geq \varepsilon |G_A^*|$ we have
$$\frac{|G_A \cap H_A^*|}{|H_A^*|} = d_A(G) \pm \varepsilon.$$

We say G is ε-*regular* if G_A is ε-regular for every $A \in \binom{[r]}{\leq k}$.

The following lemma states that the restriction of any regular and dense k-partite k-complex to large subsets of its vertex classes is also regular and dense (it is a weakened version of [**22**, Theorem 6.18]).

LEMMA 6.1. *(Regular restriction) Suppose that $1/n \ll \varepsilon \ll c, 1/k$. Let $\mathcal{Q}$ partition a set X into $X_1, \ldots, X_k$, and G be an ε-regular $\mathcal{Q}$-partite k-complex on X with $d(G) \geq c$. Then for any subsets $X_1' \subseteq X_1, \ldots, X_k' \subseteq X_k$ each of size at least $\varepsilon^{1/2k} n$, the restriction $G' = G[X_1' \cup \cdots \cup X_k']$ is $\sqrt{\varepsilon}$-regular with $d(G') \geq d(G)/2$ and $d_{[k]}(G') \geq d_{[k]}(G)/2$.*

6.2. The Regular Approximation Lemma

Roughly speaking, hypergraph regularity theory shows that an arbitrary k-graph can be split into pieces, each of which forms the 'top level' of a regular k-complex. To describe the splitting, we require the following definitions. Let $\mathcal{Q}$ partition a set X into r parts $X_1, \ldots, X_r$. A $\mathcal{Q}$-*partition k-system* on X consists of a partition $\mathcal{C}_A$ of $K_A(X)$ for each $A \in \binom{[r]}{\leq k}$. We refer to the parts of each $\mathcal{C}_A$ as *cells*, and to the cells of $\mathcal{C}_{\{i\}}$ for each $i \in [k]$ as the *clusters* of P. Observe that the clusters of P form a partition of X which refines $\mathcal{Q}$. For any $\mathcal{Q}$-partite set $S \subseteq X$ with $|S| \leq k$, we write C_S for the set of all edges of $K_{i(S)}(X)$ lying in the same cell of P as S. We write $C_{S\leq}$ for the $\mathcal{Q}$-partite k-system with vertex set X and edge set $\bigcup_{S'\subseteq S} C_{S'}$.

Let P be a $\mathcal{Q}$-partition k-system on X. We say that P is a $\mathcal{Q}$-*partition k-complex* if P satisfies the additional condition that whenever edges $S, S' \in K_A(X)$ lie in the same cell of $\mathcal{C}_A$, the edges S_B, S_B' of $K_B(X)$ lie in the same cell of $\mathcal{C}_B$ for any $B \subseteq A$. Note that then $C_{S\leq}$ is a k-complex. We say that P is *vertex-equitable* if every cluster of P has the same size. We say that P is a-*bounded* if for every $A \in \binom{[r]}{\leq k}$ the partition $\mathcal{C}_A$ partitions $K_A(X)$ into at most a cells. We say that P is ε-*regular* if $C_{S\leq}$ is ε-regular for every $\mathcal{Q}$-partite $S \subseteq X$ with $|S| \leq k$.

Now instead let P be a $\mathcal{Q}$-partition $(k-1)$-complex on X. Then P naturally induces a $\mathcal{Q}$-partition k-complex P' on X. Indeed, for any $A \in \binom{[r]}{k}$ we say that $S, S' \in K_A(X)$ are *weakly equivalent* if for any strict subset $B \subset A$ we have that S_B and S_B' lie in the same cell of $\mathcal{C}_B$; this forms an equivalence relation on $K_A(X)$. Then for each $A \in \binom{[r]}{\leq k-1}$ the partition $\mathcal{C}_A'$ of P' is identical to the partition $\mathcal{C}_A$ of P, and for each $A \in \binom{[r]}{k}$ the partition $\mathcal{C}_A'$ of P' has the equivalence classes of the weak equivalence relation as its cells. We refer to P' as the $\mathcal{Q}$-partition k-complex generated from P by weak equivalence. Note that if P is a-bounded then P' is

a^k-bounded, as for each $A \in \binom{[r]}{k}$ we have that $K_A(X)$ is divided into at most a^k cells by weak equivalence. Now let G be a $\mathcal{Q}$-partite k-graph on X. We denote by $G[P]$ the $\mathcal{Q}$-partition k-complex formed by using weak equivalence to refine the partition $\{G_A, K_A(X) \setminus G_A\}$ of $K_A(X)$ for each $A \in \binom{[r]}{k}$, i.e. two edges of G_A are in the same cell if they are weakly equivalent, and similarly for two k-sets in $K_A(X) \setminus G_A$. Together with P, this yields a partition k-complex which we denote by $G[P]$. If $G[P]$ is ε-regular then we say that G is *perfectly ε-regular with respect to P*.

We use the following form of hypergraph regularity due to Rödl and Schacht [50] (it is a slight reformulation of their result). It states that any given k-graph H can be approximated by another k-graph G that is regular with respect to some partition $(k-1)$-complex P. It is convenient to consider the k-graph G, but we take care not to use any edges in $G \setminus H$, to ensure that every edge we use actually lies in the original k-graph H. There are various other forms of the regularity lemma for k-graphs which give information on H itself, but these do not have the hierarchy of densities necessary for the application of the blow-up lemma (see [22] for discussion of this point). In the setting of the theorem, we say that G and H are ν-*close* (with respect to $\mathcal{Q}$) if $|G_A \triangle H_A| < \nu|K_A(V)|$ for every $A \in \binom{[r]}{k}$.

THEOREM 6.2. *(Regular Approximation Lemma) Suppose $1/n \ll \varepsilon \ll 1/a \ll \nu, 1/r, 1/k$ and $a!r|n$. Let V be a set of n vertices, $\mathcal{Q}$ be an balanced partition of V into r parts, and H be a $\mathcal{Q}$-partite k-graph on V. Then there is an a-bounded ε-regular vertex-equitable $\mathcal{Q}$-partition $(k-1)$-complex P on V, and an $\mathcal{Q}$-partite k-graph G on V, such that G is ν-close to H and perfectly ε-regular with respect to P.*

6.3. The hypergraph blowup lemma

While hypergraph regularity theory is a relatively recent development, still more recent is the hypergraph blow-up lemma due to Keevash [22], which makes it possible to apply hypergraph regularity theory to embeddings of spanning subcomplexes. Indeed, it is similar to the blow-up lemma for graphs, insomuch as it states that by deleting a small number of vertices from a regular k-complex we may obtain a *super-regular* complex, in which we can find any spanning subcomplex of bounded maximum degree. However, unlike the graph case, the definition of super-regularity for complexes is extremely technical, so it is more convenient to work with a formulation using robustly universal complexes. In essence, a complex J' is robustly universal if even after the deletion of many vertices (with certain conditions), we may find any spanning subcomplex of bounded degree within the complex J that remains. The formal definition is as follows (we have simplified it by removing the option of 'restricted positions', which are not required in this paper).

DEFINITION 6.3 (Robustly universal). *Suppose that V' is a set of vertices, $\mathcal{Q}$ is a partition of V' into k parts $V_1' \cup \cdots \cup V_k'$, and J' is a $\mathcal{Q}$-partite k-complex on V' with $J'_{\{i\}} = V_i'$ for each $i \in [k]$. Then we say that J' is c-robustly D-universal if whenever*

(i) *$V_j \subseteq V_j'$ are sets with $|V_j| \geq c|V_j'|$ for each $j \in [k]$, such that writing $V = \bigcup_{j \in [k]} V_j$ and $J = J'[V]$ we have $|J_k(v)| \geq c|J'_k(v)|$ for any $j \in [k]$ and $v \in V_j$, and*

(ii) L is a k-partite k-complex of maximum vertex degree at most D whose vertex classes $U_1, \ldots, U_k$ satisfy $|U_j| = |V_j|$ for each $j \in [k]$,

then J contains a copy of L, in which for each $j \in [k]$ the vertices of U_j correspond to the vertices of V_j.

The following version of the hypergraph blow-up lemma states that we may obtain a robustly universal complex from a regular complex by deleting a small number of vertices (it is a special case of [**22**, Theorem 6.32]). After applying Theorem 6.2, we regard $Z = G \setminus H$ as the 'forbidden' edges of G; so with this choice of Z in Theorem 6.4, the output $G' \setminus Z'$ is a subgraph of H.

THEOREM 6.4. *(Blow-up Lemma)* Suppose $1/n \ll \varepsilon \ll d^* \ll d_a \ll \theta \ll d, c, 1/k, 1/D, 1/C$. Let V be a set of vertices, $\mathcal{Q}$ be a partition of V into k parts $V_1, \ldots, V_k$ with $n \leq |V_j| \leq Cn$ for each $j \in [k]$, and G be an ε-regular $\mathcal{Q}$-partite k-complex on V such that $|G_{\{j\}}| = |V_j|$ for each $j \in [k]$, $d_{[k]}(G) \geq d$ and $d(G) \geq d_a$. Suppose $Z \subseteq G_k$ satisfies $|Z| \leq \theta |G_k|$. Then we can delete at most $2\theta^{1/3}|V_j|$ vertices from each V_j to obtain $V' = V_1' \cup \cdots \cup V_k'$, $G' = G[V']$ and $Z' = Z[V']$ such that
  (i) $d(G') > d^*$ and $|G'(v)_k| > d^*|G_k'|/|V_i'|$ for every $v \in V_i'$, and
  (ii) $G' \setminus Z'$ is c-robustly D-universal.

We will apply the blow-up lemma in conjunction with the following lemma for finding perfect matchings in subcomplexes of robustly universal complexes. The set X forms an 'ideal' for the perfect matching property, in that *any* extension W of X with parts of equal size contains a perfect matching. We will see later that a random choice of X has this property with high probability.

LEMMA 6.5. *Let $\mathcal{Q}$ partition a set U into $U_1, \ldots, U_k$, and G be a $\mathcal{Q}$-partite k-complex on U which is c-robustly 2^k-universal. Suppose we have $X_j \subseteq U_j$ for each $j \in [k]$ such that*
  (i) $|X_j| \geq c|U_j|$ for each $j \in [k]$, and
  (ii) $|G_k[X \cup \{v\}](v)| \geq c|G_k(v)|$ for any $v \in U$, where $X := X_1 \cup \cdots \cup X_k$.

Then for any sets W_j with $X_j \subseteq W_j \subseteq U_j$ for $j \in [k]$ and $|W_1| = \cdots = |W_k|$, writing $W = W_1 \cup \cdots \cup W_k$, the k-complex $G[W]$ contains a perfect matching,

PROOF. By (i) we have $|W_j| \geq |X_j| \geq c|U_j|$ for each j, and by (ii) we have

$$|G_k[W](v)| \geq |G_k[X \cup \{v\}](v)| \geq c|G_k(v)|$$

for every $v \in W$. So by definition of a c-robustly 2^k-universal complex, G contains any k-partite k-complex on W with maximum vertex degree at most 2^k. In particular, this includes the complex obtained by the downward closure of a perfect matching in $K_{[k]}[W]$. □

6.4. Reduced k-systems

Now we introduce and develop the theory of reduced k-systems, whose role in the k-system setting is analogous to that of reduced graphs in the graph setting. Informally speaking, an edge of the reduced system represents a set of clusters, which is dense in the original k-system J, and sparse in the forbidden k-graph Z. While this does not contain enough information for embeddings of general k-graphs, it is sufficient for matchings, which is our concern here. Note that this definition includes a partition $(k-1)$-complex P, which will be obtained from the regularity lemma, but only the first level of this complex (the partition of X into clusters) is used in the definition.

DEFINITION 6.6 (Reduced k-system). *Let $\mathcal{Q}$ partition a set X of size n into parts of equal size, and let J be a $\mathcal{Q}$-partite k-system, Z a $\mathcal{Q}$-partite k-graph, and P a vertex-equitable $\mathcal{Q}$-partition $(k-1)$-complex on X. Suppose $\nu \in \mathbb{R}$ and $\mathbf{c} = (c_1, \ldots, c_k) \in \mathbb{R}^k$. We define the* reduced k-system $R := R_{P\mathcal{Q}}^{JZ}(\nu, \mathbf{c})$ *as follows.*

Let $V_1, \ldots, V_m$ be the clusters of P, and let n_1 denote their common size. The vertex set of R is $[m]$, where vertex i corresponds to cluster V_i of P. Any partition $\mathcal{P}$ of X which is refined by the partition of X into clusters of P naturally induces a partition of $[m]$, which we denote by $\mathcal{P}_R$: i and j lie in the same part of $\mathcal{P}_R$ if and only if V_i and V_j are subsets of the same part of $\mathcal{P}$. Note that $\mathcal{Q}_R$ partitions $[m]$ into parts of equal size.

The edges of R are defined as $\emptyset$, and those $S \in \binom{[m]}{j}$ for each $j \in [k]$ such that S is $\mathcal{Q}_R$-partite, $|J[\bigcup_{i \in S} V_i]| \geq c_j n_1^j$, and for any $S' \subseteq S$ of size j', at most $\nu^{2-j'} n_1^{j'} n^{k-j'}$ edges of Z intersect each member of $\{V_i : i \in S'\}$.

Ideally, we would like the reduced k-system of a k-complex J to be itself a k-complex. However, it does not appear to be possible to define a reduced k-system that both accomplishes this and inherits a minimum degree condition similar to that of J. Instead, we have a weaker property set out by the following result, which shows that any subset of an edge of a reduced k-system of J is an edge of another reduced k-system of J, where the latter has weaker density parameters.

LEMMA 6.7. *Let $\mathcal{Q}$ partition a set X into parts of equal size, J be a $\mathcal{Q}$-partite k-system, Z a $\mathcal{Q}$-partite k-graph, and P a vertex-equitable $\mathcal{Q}$-partition $(k-1)$-complex on X, with clusters of size n_1. Suppose $\mathbf{c}, \mathbf{c}' \in \mathbb{R}^k$ with $0 \leq c_i' \leq c_j$ for all $i, j \in [k]$ with $i \leq j$. Then for any $e \in R := R_{P\mathcal{Q}}^{JZ}(\nu, \mathbf{c})$ and $e' \subseteq e$ we have $e' \in R' := R_{P\mathcal{Q}}^{JZ}(\nu, \mathbf{c}')$.*

PROOF. Since $e \in R$, e is $\mathcal{Q}_R$-partite, and so e' is $\mathcal{Q}_R$-partite, which is identical to being $\mathcal{Q}_{R'}$-partite. Next, since $e \in R$, for any $e'' \subseteq e$ of size j at most $\nu^{2-j} n_1^j n^{k-j}$ edges of Z intersect each member of $\{V_i : i \in e''\}$. In particular, this property holds for any $e'' \subseteq e'$. Finally, since $e \in R$, we have $|J[\bigcup_{i \in e} V_i]| \geq c_{|e|} n_1^{|e|}$. Since J is a k-complex and $c'_{|e'|} \leq c_{|e|}$, it follows that $|J[\bigcup_{i \in e'} V_i]| \geq c'_{|e'|} n_1^{|e'|}$. □

The next lemma shows that when the k-graph Z is sparse, we do indeed have the property mentioned above, namely that the minimum degree sequence of the original k-system J is 'inherited' by the reduced k-system. We work in the more general context of minimum F-degree sequences defined in Chapter 5.

LEMMA 6.8. *Suppose that $1/n \ll 1/h, \nu \ll c_k \ll \cdots \ll c_1 \ll \alpha, 1/k, 1/r$. Let X be a set of rn vertices, $\mathcal{P}$ partition X into r parts $X_1, \ldots, X_r$ of size n, and $\mathcal{Q}$ refine $\mathcal{P}$ into h parts of equal size. Let J be a $\mathcal{Q}$-partite $(k-1)$-system on X, Z a $\mathcal{Q}$-partite k-graph on X with $|Z| \leq \nu n^k$, and P a vertex-equitable $\mathcal{Q}$-partition $(k-1)$-complex on X with clusters $V_1, \ldots, V_{rm}$ of size n_1. Suppose also that J and Z are $\mathcal{P}F$-partite for some allocation F, and that $\delta_0^F(J) = n$. Then, with respect to $\mathcal{P}_R$, the reduced k-system $R := R_{P\mathcal{Q}}^{JZ}(\nu, \mathbf{c})$ satisfies*

$$\delta^F(R) \geq \left((1 - k\nu^{1/2})m, \left(\frac{\delta_1^F(J)}{n} - \alpha\right)m, \ldots, \left(\frac{\delta_{k-1}^F(J)}{n} - \alpha\right)m\right).$$

PROOF. Let $U_1, \ldots, U_r$ be the parts of $\mathcal{P}_R$ corresponding to $X_1, \ldots, X_r$ respectively. Fix any $f \in F$, $j \in [k-1]$ and $S = \{u_1, \ldots, u_j\} \in R_j$ with $u_i \in U_{f(i)}$ for $i \in [j]$. Then S is $\mathcal{Q}_R$-partite and $\mathcal{P}_R F$-partite, and $|J[V_S]| \geq c_j n_1^j$. For any edge $e \in J[V_S]$, we may write $e = \{v_1, \ldots, v_j\}$ with $v_i \in X_{f(i)}$ for $i \in [j]$. There are at least $\delta_j^F(J)$ vertices $v_{j+1} \in X_{f(j+1)}$ such that $\{v_1, \ldots, v_{j+1}\} \in J$, and of these at most jrn/h belong to the same part of $\mathcal{Q}$ as one of $v_1, \ldots, v_j$. Thus we obtain at least $|J[V_S]|(\delta_j^F(J) - jrn/h)$ edges in sets $J[V_T]$, $T \in \mathcal{T}$, where $\mathcal{T}$ denotes the collection of $\mathcal{Q}_R$-partite sets $T = S \cup \{u\}$ for some $u \in U_{f(j+1)} \setminus S$. At most $mc_{j+1}n_1^{j+1} = c_{j+1}nn_1^j$ of these edges belong to sets $J[V_T]$ of size less than $c_{j+1}n_1^{j+1}$, so there are at least $\delta_j^F(J)m/n - \alpha m/2$ sets $T \in \mathcal{T}$ with $|J[V_T]| \geq c_{j+1}n_1^{j+1}$.

Such a set T is an edge of R, unless for some $T' \subseteq T$ there are more than $\nu^{2-2^{j'-1}} n_1^{j'+1} n^{k-j'-1}$ edges of Z which intersect each part of $V_{T'}$, where $|T'| = j'+1$ for some $0 \leq j' \leq j$; in this case we say that T is T'-bad. Note that such a set T' is not contained in S, otherwise S would be T'-bad, contradicting $S \in R$. So we can write $T' = S' \cup \{u\}$, where $S' \subseteq S$ and $T = S \cup \{u\}$. For fixed $S' \subseteq S$, there can be at most $\nu^{2-2^{j'-1}} m$ such vertices u, otherwise S would be S'-bad, contradicting $S \in R$. Summing over all $S' \subseteq S$ we find that there are at most $2^j \nu^{2-2^{-k}} m$ vertices $u \in [U_{f(j+1)}]$ such that $T = S \cup \{u\}$ is T'-bad for some $T' \subseteq T$. It follows that

$$\delta_j^F(R) \geq \delta_j^F(J)m/n - \alpha m/2 - 2^j \nu^{2-2^{-k}} m \geq \delta_j^F(J)m/n - \alpha m.$$

It remains to show that $\delta_0^F(R) \geq (1 - k\nu^{1/2})m$, i.e. for each $i \in [r]$ which lies in the image of some $f \in F$, the set $\{u\}$ is an edge of R for all but at most $k\nu^{1/2}m$ vertices $u \in U_i$. For any $u \in U_i$, the set $\{u\}$ is $\mathcal{Q}_R$-partite, and since $\delta_0^f(J) = n$ we have $|J[V_u]| = n_1$. So $\{u\}$ is an edge of R unless more than $\nu^{1/2} n_1 n^{k-1}$ edges of Z intersect V_u; since $|Z| \leq \nu n^k$ this is true for at most $k\nu^{1/2}m$ vertices u, as required. $\square$

The next proposition shows that the k-edges of the reduced k-system are useful, in that the corresponding clusters contain a sub-k-graph of J_k to which the blow-up lemma can be applied; here we note that the complex G' obtained in the proposition meets the conditions of Theorem 6.4 (with $\theta = \nu^{1/3}$ and d replaced by $d/2$).

PROPOSITION 6.9. *Suppose that $1/n \ll \varepsilon \ll d_a \ll 1/a \ll \nu \ll d \ll 1/k$. Let X be a set of n vertices, $\mathcal{Q}$ partition X into parts of equal size, G and Z be $\mathcal{Q}$-partite k-graphs on X, and P be a vertex-equitable a-bounded $\mathcal{Q}$-partition $(k-1)$-complex on X, with clusters of size n_1. Suppose that G is perfectly ε-regular with respect to P, and $U_1, \ldots, U_k$ are clusters of P such that $U := U_1 \cup \cdots \cup U_k$ satisfies $|Z[U]| \leq \nu n_1^k$ and $|G[U]| \geq dn_1^k$. Let $\mathcal{P}$ denote the partition of U into parts $U_1, \ldots, U_k$. Then there exists a $\mathcal{P}$-partite k-complex G' on U such that $G'_k \subseteq G$, G' is ε-regular, $d_{[k]}(G') \geq d/2$, $d(G') \geq d_a$, and $Z' = Z \cap G'_k$ has $|Z'| \leq \nu^{1/3}|G'_k|$.*

PROOF. To find G' we select a suitable cell of P^*, which we recall is the a^k-bounded $\mathcal{Q}$-partition k-complex formed from P by weak equivalence. Note that one of the partitions forming P^* is a partition of $K_{[k]}(U)$ into cells $C_1, \ldots, C_s$, where $s \leq a^k$. So at most $dn_1^k/3$ edges of $K_{[k]}(U)$ lie within cells C_i such that $|C_i| \leq dn_1^k/(3a^k)$. Also, since $|Z[U]| \leq \nu n_1^k$, at most $\nu^{1/2} n_1^k$ edges of $K_{[k]}(U)$ lie within cells C_i such that $|Z \cap C_i| \geq \nu^{1/2}|C_i|$. Then, since $|G[U]| \geq dn_1^k$, at least $dn_1^k/2$ edges of G must lie within cells C_i with $|C_i| > dn_1^k/(3a^k)$ and $|Z \cap C_i| < \nu^{1/2}|C_i|$.

By averaging, there must exist such a cell C_i that also satisfies $|G \cap C_i| > d|C_i|/2$. Fix such a choice of C_i, which we denote by C.

We now define G' to be the complex with top level $G'_k = G \cap C$ and lower levels $G'_{<k} = C_{<k} = \bigcup_{S' \subset S} C_{S'}$, where $S \in C$. Then $G'_k \subseteq G$, and G' is ε-regular, as G is perfectly ε-regular with respect to P. Furthermore, we have

$$d_{[k]}(G') = \frac{|G'_{[k]}|}{|(G')^*_{[k]}|} = \frac{|G \cap C|}{|C|} > d/2$$

and

$$d(G') = \frac{|G'_{[k]}|}{|K_{[k]}(U)|} = \frac{|G \cap C|}{|C|} \cdot \frac{|C|}{n_1^k} > \frac{d^2}{6a^k} > d_a.$$

Finally, since $|G'_k| = |G \cap C| > d|C|/2$ and $|Z'| \leq |Z \cap C| < \nu^{1/2}|C|$, we have $|Z'| < \frac{2\nu^{1/2}}{d}|G'_k| < \nu^{1/3}|G'_k|$. □

Finally, we need the following lemma, which states (informally) that index vectors where the original k-system J is dense and the forbidden k-graph Z is sparse are inherited as index vectors where the reduced k-system R is dense. If J is a k-system on X, and $\mathcal{P}$ partitions X, then we denote by $J_\mathbf{i}$ the set of edges in J with index $\mathbf{i}$ with respect to $\mathcal{P}$.

LEMMA 6.10. *Suppose that* $1/n \ll \nu \ll c_k \ll \cdots \ll c_1 \ll \mu, 1/k$. *Let* X *be a set of* n *vertices*, $\mathcal{Q}$ *partition* X *into parts of equal size*, J *be a* $\mathcal{Q}$-*partite* k-*system*, Z *a* $\mathcal{Q}$-*partite* k-*graph with* $|Z| \leq \nu n^k$, *and* P *a vertex-equitable* $\mathcal{Q}$-*partition* $(k-1)$-*complex on* X, *with clusters* $V_1, \ldots, V_m$ *of size* n_1. *Let* $R := R^{JZ}_{P\mathcal{Q}}(\nu, \mathbf{c})$ *be the reduced* k-*system. Then for any partition* $\mathcal{P}$ *of* $V(J)$ *which is refined by the partition of* $V(J)$ *into clusters of* P *we have the following property:*

(**Inheritance of index vectors**): *If* $|J_\mathbf{i}| \geq \mu n^p$, *where* $p = |\mathbf{i}|$, *then* $|R_\mathbf{i}| \geq \mu m^p/2$, *where index vectors are taken with respect to* $\mathcal{P}$ (*for* J) *and* $\mathcal{P}_R$ (*for* R).

PROOF. Let $\mathcal{B}$ be the set of $\mathcal{Q}_R$-partite sets S of size p with $S \notin R_p$. We estimate the number of edges contained in all p-graphs $J_p[V_S]$ with $S \in \mathcal{B}$. There are two reasons for which we may have $S \notin R_p$. The first is that $|J_p[V_S]| < c_p n_1^p$; this gives at most $c_p n^p$ edges in total. The second is that S is S'-bad for some $S' \subseteq S$, i.e. more than $\nu^{2-j'} n_1^{j'} n^{k-j'}$ edges of Z intersect each member of $\{V_i : i \in S'\}$, where $j' = |S'|$. For any given $j' \leq p$, there can be at most $\binom{p}{j'} \nu^{1/2} m^p$ such sets S, otherwise we would have at least $\binom{p}{j'} \nu^{1/2} m^{j'}$ sets S' of size j' for which more than $\nu^{1/2} n_1^{j'} n^{k-j'}$ edges of Z intersect each member of $\{V_i : i \in S'\}$, contradicting $|Z| \leq \nu n^k$. Summing over $j' \leq p$, this gives at most $2^p \nu^{1/2} m^p n_1^p \leq c_p n^p$ edges in total. Thus there are at most $2c_p n^p$ edges contained in all p-graphs $J_p[V_S]$ with $S \in \mathcal{B}$. Now, if at least μn^p edges $e \in J_p$ have $\mathbf{i}_\mathcal{P}(e) = \mathbf{i}$, then at least $\mu n^p/2$ of these edges lie in some $J_p[V_S]$ with $S \in R_p$. For each such S we have $\mathbf{i}_{\mathcal{P}_R}(S) = \mathbf{i}$ and $|J_p[V_S]| \leq n_1^p$. Thus there are at least $\mu m^p/2$ edges $S \in R_p$ with $\mathbf{i}_{\mathcal{P}_R}(S) = \mathbf{i}$. □

6.5. Proof of Lemma 5.5

In this section it is more convenient to use the (much simpler) Weak Regularity Lemma, in the context of a simultaneous regularity partition for several hypergraphs. Suppose that $\mathcal{P}$ partitions a set V into r parts $V_1, \ldots, V_r$ and G is a $\mathcal{P}$-partite k-graph on V. For $\varepsilon > 0$ and $A \in \binom{[r]}{k}$, we say that the k-partite sub-k-graph G_A is (ε, d)-*vertex-regular* if for any sets $V_i' \subseteq V_i$ with $|V_i'| \geq \varepsilon |V_i|$ for $i \in A$, writing $V' = \bigcup_{i \in A} V_i'$, we have $d(G_A[V']) = d \pm \varepsilon$. We say that G_A is ε-*vertex-regular* if it is (ε, d)-vertex-regular for some d. The following lemma has essentially the same proof as that of the Szemerédi Regularity Lemma [55], namely iteratively refining a partition until an 'energy function' does not increase by much; in this case the energy function is the sum of the mean square densities of the k-graphs with respect to the partition.

THEOREM 6.11. *(Weak Regularity Lemma) Suppose that $1/n \ll 1/m \ll \varepsilon \ll 1/t, 1/r \leq 1/k$. Suppose that $G^1, \ldots, G^t$ are k-graphs on a set V of n vertices. Then there is a partition $\mathcal{P}$ of V into $m' \leq m$ parts, such that there is some n_0 so that each part of $\mathcal{P}$ has size n_0 or $n_0 + 1$, and for each $i \in [t]$, all but at most εn^k edges of G^i belong to ε-vertex-regular k-partite sub-k-graphs.*

We also require the fact that regularity properties are inherited by random subsets with high probability. We use the formulation given by Czygrinow and Nagle [9, Theorem 1.2] (a similar non-partite statement was proved earlier by Mubayi and Rödl [41]).

THEOREM 6.12. *Suppose that $1/n \ll 1/s, \varepsilon, c \ll \varepsilon', 1/k$, that $\mathcal{P}$ partitions a set V into k parts $V_1, \ldots, V_k$ of size at least n, and G is an (ε, d)-vertex-regular k-partite k-graph on V. Suppose that $s_i \geq s$ for $i \in [k]$, and sets $S_i \subseteq V_i$ of size s_i are independently chosen uniformly at random for $i \in [k]$. Let $S = \bigcup_{i \in [k]} S_i$. Then $G[S]$ is (ε', d)-vertex-regular with probability at least $1 - e^{-c \min_i s_i}$.*

Here and later in the paper we will need the following inequality known as the Chernoff bound, as applied to binomial and hypergeometric random variables. We briefly give the standard definitions. The binomial random variable with parameters (n, p) is the sum of n independent copies of the $\{0, 1\}$-valued variable A with $\mathbb{P}(A = 1) = p$. The hypergeometric random variable X with parameters (N, m, n) is defined as $X = |T \cap S|$, where $S \subseteq [N]$ is a fixed set of size m, and $T \subseteq [N]$ is a uniformly random set of size n. If $m = pN$ then both variables have mean pn.

LEMMA 6.13 ([20, Corollary 2.3 and Theorem 2.10]). *Suppose X has binomial or hypergeometric distribution and $0 < a < 3/2$. Then $\mathbb{P}(|X - \mathbb{E}X| \geq a\mathbb{E}X) \leq 2e^{-\frac{a^2}{3}\mathbb{E}X}$.*

Loosely speaking, the proof of Lemma 5.5 proceeds as follows. We divide the edges of M into 'blocks', where each block contains one edge of each index vector in $I(F)$. Since M α'-represents F, we can do this so that only a few edges of M are not included in some block (we discard these edges). We then choose M' to consist of the edges in the blocks covering u and v, as well as the edges in a further $n'/b - 2$ blocks chosen uniformly at random, where b is the number of edges in each block. The set V' of vertices covered by M' then includes n' vertices from each part of $\mathcal{P}$, and induces a matched k-system (J', M'). Also, a fairly straightforward application of the Chernoff bound implies that with high probability the induced k-system (J', M') inherits the minimum degree sequence of J (with slightly greater error terms) giving condition (i).

The main difficulty is therefore to show that condition (ii) also holds with high probability, for which we use weak hypergraph regularity. This argument is similar in spirit to methods used for property-testing (see e.g. [51]), though it does not follow directly from them. To see the main ideas of this part of the proof, suppose that we apply Theorem 6.11 to partition each V_j into m equally-sized clusters, and assume for the sake of simplicity that any k clusters $U_1, \ldots, U_k$ from different parts of $\mathcal{P}$ induce a vertex-regular sub-k-graph of J_k (in reality this will only be true of most such sets of k clusters). With high probability our randomly chosen V' will include approximately the same number of vertices from each cluster, and by Theorem 6.12 the induced subclusters $U'_1, \ldots, U'_k$ of any k clusters $U_1, \ldots, U_k$ from different parts of $\mathcal{P}$ will also induce a vertex-regular subgraph of G with similar density. Now suppose some set $S' \subseteq V'$ contains $\lfloor pn'/k \rfloor$ vertices from each V_j. Then S' must have not-too-small intersection with approximately pm/k subclusters from each part of $\mathcal{P}$. But if S' has not-too-small intersection with each of $U'_1, \ldots, U'_k$, then the vertex-regularity of $U'_1, \ldots, U'_k$ implies the density of the induced sub-k-graph $G[S' \cap (U'_1 \cup \cdots \cup U'_k)]$ is approximately equal to the density of $G[U'_1 \cup \cdots \cup U'_k]$, which in turn is approximately equal to the density of $G[U_1 \cup \cdots \cup U_k]$. It follows that the density of $G[S']$ is approximately equal to the density of $G[S]$, where S is the union of the clusters whose subclusters have not-too-small intersection with S'. However, S must contain close to pn/k vertices in each part of $\mathcal{P}$, and so condition (ii) on J implies that $G[S']$ is not too small, from which we deduce that condition (ii) on J' holds for (our arbitrary choice of) S'.

In the above sketch we glossed over several significant difficulties. Perhaps foremost among these is the fact that Theorem 6.12 requires the choice of vertices to be uniformly random, which is certainly not the case here. To resolve this difficulty, rather than apply Theorem 6.11 directly to G, we instead apply it to a family of auxiliary k-graphs J^{TC}, each of which has the blocks of edges of M' as vertices. Ignoring the blocks which cover the vertices of u and v, our random choice is then to take a subset of the vertex set of each J^{TC} uniformly at random, so we may indeed apply Theorem 6.12.

We now give the full details of the the proof of Lemma 5.5. We introduce new constants $\varepsilon, \varepsilon', c, d, m$ such that $1/n' \ll 1/m \ll \varepsilon, c \ll \varepsilon' \ll d \ll \alpha'$. Let e_u and e_v be the edges of M containing u and v respectively. Since M α'-represents F, we can partition M as $M_0 \cup M_1$, where M_1 is F-balanced, $\{e_u, e_v\} \subseteq M_1$ and $|M_0| \leq \alpha'|M|$. Letting V_0 be the set of vertices covered by M_0, we have $|V_0| = k|M_0| \leq \alpha' rn$. Since $|M| = rn/k$ we can write $|M_1| = rn_1/k$ with $n_1 \geq (1-\alpha')n$. By uniformity of F, M_1 covers n_1 vertices in each part of $\mathcal{P}$. Let $I := I(F)$ (recall that if $I(F)$ is a set then this is the set $\{\mathbf{i}(f) : f \in F\}$). So $|I| = |F|/k!$. We partition M_1 into blocks, where each block contains one edge e' with $\mathbf{i}(e') = \mathbf{i}$ for each $\mathbf{i} \in I$. So each block consists of $|I|$ edges, and uses $b := k|I|/r$ vertices from each part. We arbitrarily label the blocks as B_i, $i \in [n_B]$, where $n_B := |M_1|/|I| = n_1/b$, and for each $j \in [r]$ we let B_{ij} be the vertices in V_j used by edges of B_i, which we arbitrarily label by $[b]$. For $x \in [n_B]$, $t \in [r]$, $c \in [b]$ we let v_{xtc} denote vertex c in B_{xt}. We construct M' as a union of blocks. We start by taking two distinct blocks B_{x_u} and B_{x_v}, where $e_u \in B_{x_u}$, and either $e_v \in B_{x_v}$ or $e_v \in B_{x_u}$ and B_{x_v} is arbitrary. Then we let X be a subset of $[n_B] \setminus \{x_u, x_v\}$ of size $n'/b - 2$ chosen uniformly at random, and let $X' = X \cup \{x_u, x_v\}$, so $|X'| = n'/b$. Take

$$M' = \bigcup_{x \in X'} B_x, \quad V' = \bigcup_{e \in M'} e, \quad J' = J[V'] \text{ and } V'_i = V_i \cap V' \text{ for each } i \in [r].$$

Then we have $u, v \in V'_a$, $|V'_1| = \cdots = |V'_r| = n'$, and (J', M') is a matched $\mathcal{P}F$-partite k-system on V'.

Fix $f \in F$ and consider the minimum f-degree sequence $\delta^f(J')$. We clearly have $\delta^f_0(J') = n'$, as for any $v \in V'$ we have $\{v\} \in J$, and so $\{v\} \in J'$. Now fix any $e = \{v_1, \ldots, v_j\} \in J$ for some $j \in [k-1]$ such that $v_i \in V_{f(i)}$ for $i \in [j]$. For each $i \in [j]$ let B_{d_i} be the block containing v_i, so $\mathcal{B}_e := B_{x_u} \cup B_{x_v} \cup \bigcup B_{d_i}$ is the union of $\ell \leq j+2$ distinct blocks. Let A be the set of vertices covered by edges in $\mathcal{B}_e$. Write $q = f(j+1)$, and for $c \in [b]$, let $N(e)_c$ be the set of vertices $v \in V_q \setminus (V_0 \cup A)$ such that $e \cup \{v\} \in J$ and $v = v_{xqc}$ for some $x \in [n_B]$. Then $\sum_{c \in [b]} |N(e)_c| \geq \left(\frac{k-j}{k} - \alpha\right)n - \alpha' rn - \ell br$ by assumption on $\delta^f(J)$. Writing $|N(e)_c| = \theta_c n_1$, and $C = \{c \in [b] : \theta_c > 2\varepsilon\}$ we have

$$\sum_{c \in C} \theta_c \geq 1 - j/k - \alpha - 2\alpha' r - 2b\varepsilon.$$

Note that, conditioning on the event $e \in J'$, the random variable $|N(e)_c \cap V'_q|$ is hypergeometric with mean $\frac{n'/b-\ell}{n_1/b-\ell}|N(e)_c| > (\theta_c - \varepsilon)n'$ for each $c \in C$, so the Chernoff bound (Lemma 6.13) gives $|N(e)_c \cap V'_q| > (\theta_c - 2\varepsilon)n'$ with probability at least $1 - 2e^{-\varepsilon^3 n'/3}$. On these events we have

$$m^f(e) := \sum_{c \in [b]} |N(e)_c \cap V'_q| > \sum_{c \in C} (\theta_c - 2\varepsilon)n' > (1 - j/k - 2\alpha)n'.$$

So, letting $Z(e, f)$ denote the event that $m^f(e) \leq (1 - j/k - 2\alpha)n'$, and no longer conditioning on $e \in J'$, we have

$$\mathbb{P}(e \in J' \text{ and } Z(e, f)) = \mathbb{P}(e \in J')\mathbb{P}(Z(e, f) \mid e \in J') \leq \left(\frac{n'/b - \ell}{n/b - \ell}\right)^{\ell - 2} \cdot 2e^{-\varepsilon^3 n'/3}.$$

Taking a union bound over at most $|F|(k-1)(k+1)$ choices of $f \in F$, $j \in [k-1]$ and $\ell \in [j+1]$, and at most $(n_B)^{\ell-2}(b\ell)^j \leq n^{\ell-2}(2bk)^k$ edges $e \in J_j$ such that $\mathcal{B}_e$ is the union of ℓ distinct blocks, we see that the minimum F-degree property of J' holds with high probability.

For property (ii) we use weak hypergraph regularity. Fix $p \in [k-1]$ and consider the following auxiliary $(p+1)$-graphs J^{TC} on $[n_B]$. Given $1 \leq x_1 < \cdots < x_{p+1} \leq n_B$, $T = (t_1, \ldots, t_{p+1}) \in [r]^{p+1}$ such that there is some $f \in F$ for which $f(i) = t_i$ for $i \in [p+1]$, and $C = (c_1, \ldots, c_{p+1}) \in [b]^{p+1}$, we say that $\{x_1, \ldots, x_{p+1}\} \in J^{TC}$ if and only if $\{v_{x_i t_i c_i} : i \in [p+1]\} \in J_{p+1}$. Thus each edge of J_{p+1} that has at most one vertex in any block corresponds to a unique edge in at most $(p+1)!$ of the $(p+1)$-graphs J^{TC} (these are given by holding C constant and permuting T). Observe that this accounts for most edges of J_{p+1}, as at most $br^{p+1}n^p$ edges have more than one vertex in some block. By Theorem 6.11, there is a partition $\mathcal{P}'$ of $[n_B]$ into $m' \leq m$ parts, such that there is some n_0 so that each part of $\mathcal{P}'$ has size n_0 or $n_0 + 1$, and for each T and C as above, all but at most εn^{p+1} edges of J^{TC} belong to ε-vertex-regular $\mathcal{P}'$-partite sub-$(p+1)$-graphs. Note that $m'n_0 \leq n_B \leq m'(n_0+1)$. Similarly to the previous argument, we will see that with high probability X' represents all parts of $\mathcal{P}'$ approximately equally. Indeed, fix $U \in \mathcal{P}'$, and note that, since $n_B = n_1/b$, the random variable $|X' \cap U \setminus \{x_u, x_v\}|$ is hypergeometric with mean

$$\frac{n'/b - 2}{n_1/b - 2}(n_0 \pm 2) = \frac{n'}{n_1}(n_B/m' \pm 4) = \frac{n'}{bm'} \pm 1.$$

By the Chernoff bound, we have $|X' \cap U| = (1 \pm \varepsilon)n'/bm'$ with probability at least $1 - 2e^{-\varepsilon^2 n'/6bm'}$. We restrict attention to choices of X' such that this estimate holds for all $U \in \mathcal{P}'$. Note that conditional on any specified values of $|X' \cap U|$ for $U \in \mathcal{P}'$, the choices of $X' \cap U \setminus \{x_u, x_v\}$, $U \in \mathcal{P}'$ are independent uniformly random $|X' \cap U \setminus \{x_u, x_v\}|$-sets in $U \setminus \{x_u, x_v\}$ for $U \in \mathcal{P}'$. Now consider the 'reduced' $(p+1)$-graphs R^{TC}, where $V(R^{TC}) = \mathcal{P}'$, and $E(R^{TC})$ consists of all $(p+1)$-tuples $(U_1, \ldots, U_{p+1})$ of parts of $\mathcal{P}'$, such that writing $U = \bigcup_{i \in [p+1]} U_i$, $J^{TC}[U]$ is (ε, d')-vertex-regular for some $d' \geq d$. Observe that at most $(d+\varepsilon)n^{p+1}$ edges of J^{TC} do not belong to $J^{TC}[U]$ for some such edge. Given an edge $(U_1, \ldots, U_{p+1}) \in E(R^{TC})$, and a random choice of X' such that $|X' \cap U| = (1 \pm \varepsilon)n'/bm'$ for all $U \in \mathcal{P}'$, Theorem 6.12 implies that $J^{TC}[U \cap X']$ is (ε', d')-vertex-regular with probability at least $1 - e^{-cn'/2bm'}$. We can assume that this holds for all T, C and $(U_1, \ldots, U_{p+1})$, as there are at most $(rbm)^{p+1}$ choices, so we can take a union bound.

Now consider any sets $S'_t \subseteq V'_t$ such that $|S'_t| = \lfloor pn'/k \rfloor$ for each $t \in [r]$. We need to show that there are at least $\beta'(n')^{p+1}$ edges in $J_{p+1}[S']$, where $S' := \bigcup_{t \in [r]} S'_t$. Let $S'_{tc} = \{x \in [n_B] : v_{xtc} \in S'\}$ for $t \in [r]$ and $c \in [b]$. Let $Y_{tc} = \{U \in \mathcal{P}' : |S'_{tc} \cap U| \geq \beta^2 |X' \cap U|\}$. For each $t \in [r]$ we have

$$pn'/k = |S'_t| = \sum_{c \in [b]} \sum_{U \in \mathcal{P}'} |S'_{tc} \cap U| \leq \sum_{c \in [b]} |Y_{tc}|(1+\varepsilon)n'/bm' + b\beta^2 |X'|,$$

so

$$\sum_{c \in [b]} |Y_{tc}| > \frac{pn'/k - b\beta^2 n'}{(1+\varepsilon)n'/bm'} > (p/k - 2b\beta^2)bm'.$$

Write $S_t^0 = \bigcup_{c \in [b]} \bigcup_{U \in Y_{tc}} U$. Then $|S_t^0| > (p/k - 2b\beta^2)bm'n_0 > (p/k - 3b\beta^2)n$, since $m'(n_0 + 1) \geq n_B = n/b$. Let $S_t \subseteq V_t$ be an arbitrary set of size $\lfloor pn/k \rfloor$ such that $S_t^0 \subseteq S_t$ if $|S_t^0| \leq pn/k$ or $S_t \subseteq S_t^0$ if $|S_t^0| \geq pn/k$. Let $S = S_1 \cup \cdots \cup S_r$ and $S_{tc} = \{x \in [n_B] : v_{xtc} \in S\}$ for $t \in [r]$ and $c \in [b]$. By assumption (ii), there are at least βn^{p+1} edges in $J_{p+1}[S]$. Of these edges, we discard a small number of edges that are 'bad' for one of the following reasons: at most $3rk\beta^2 n^{p+1}$ edges which are incident with $\bigcup_{t \in [r]}(S_t \setminus S_t^0)$, at most $br^{p+1}n^p$ edges which have more than one vertex in some block, and at most $(br)^{p+1}(d+\varepsilon)n^{p+1}$ edges which correspond to an edge in some J^{TC} belonging to a $(p+1)$-partite sub-$(p+1)$-graph that is not (ε, d')-vertex-regular with $d' \geq d$. This still leaves at least $(\beta - \beta^{3/2})n^{p+1}$ edges, each of which corresponds to an edge $\{x_1, \ldots, x_{p+1}\}$ in some J^{TC} such that $x_i \in S_{t_i c_i} \cap U_i$ for some $U_i \in Y_{t_i c_i}$ for $i \in [p+1]$ and $(U_i : i \in [p+1])$ is an edge of R^{TC}. Let Q^{TC} be the set of edges $(U_i : i \in [p+1])$ in R^{TC} such that $U_i \in Y_{t_i c_i}$ for $i \in [p+1]$. Then we have

$$\sum_{T,C} |Q^{TC}| \geq (\beta - \beta^{3/2})n^{p+1}/(n_0+1)^{p+1} \geq \frac{1}{2}\beta(bm')^{p+1}.$$

Recall that each edge of J with at most one vertex in any block corresponds to at most $(p+1)!$ edges in the $(p+1)$-graphs J^{TC}. So $(p+1)!|J_{p+1}[S']|$ is at least the number of edges $\{x_1, \ldots, x_{p+1}\}$ in some J^{TC} such that $x_i \in S'_{t_i c_i} \cap U_i$ for $i \in [p+1]$

where $(U_i : i \in [p+1]) \in Q^{TC}$. For each $(U_i : i \in [p+1]) \in Q^{TC}$ we have at least

$$(d - \varepsilon') \prod_{i \in [p+1]} |S'_{t_i c_i} \cap U_i| > \frac{1}{2} d(\beta^2 n'/bm')^{p+1}$$

such edges of J^{TC}, using the definition of $Y_{t_i c_i}$ and ε'-vertex-regularity. Since $\sum_{T,C} |Q^{TC}| \geq \frac{1}{2}\beta(bm')^{p+1}$ we obtain at least $\beta'(n')^{p+1}$ edges in $J_{p+1}[S'] = J'_{p+1}[S']$. □

CHAPTER 7

Matchings in k-systems

In this chapter we prove our theorems on matchings in k-systems with a minimum degree sequence condition. In fact, we prove theorems in the setting of minimum F-degree sequences, which simultaneously generalise both our non-partite and partite theorems. In the first section we prove the general form of the fractional perfect matching result. We combine this with hypergraph regularity in the second section to prove a common generalisation of Theorems 2.2 and 2.5 on almost perfect matchings. In the third section we apply transferrals to prove a common generalisation of Theorems 2.9 and 2.10 on perfect matchings; we will also see that essentially the same proof gives a common generalisation of Theorems 2.4 and 2.7.

7.1. Fractional perfect matchings

In this section we prove a lemma that generalises Lemma 3.6 to the minimum F-degree setting. Let J be a $\mathcal{P}F$-partite k-system on V, where $\mathcal{P}$ is a balanced partition of V into r parts of size n, and F is a (k, r)-uniform allocation. Recall that a fractional perfect matching in J_k is an assignment of a weight $w_e \geq 0$ to each edge $e \in J_k$ such that for any $v \in V(J_k)$ we have $\sum_{e \ni v} w_e = 1$. The lemma will show that if J satisfies our minimum F-degree condition and has no space barrier then J_k admits a fractional perfect matching (in fact, with a slightly stronger minimum degree the latter condition is not required). We actually prove something stronger, namely that J_k contains a fractional perfect matching which is F-balanced, in that $\sum_{e \in J_k: \mathbf{i}(e)=\mathbf{i}} w_e$ is constant over all $\mathbf{i} \in I(F)$; this can be seen as a fractional equivalent of an F-balanced matching as previously defined. We also say that a multiset E in J_k is F-balanced if the number of edges in E of index $\mathbf{i}$ (counted with multiplicity) is the same for any $\mathbf{i} \in I(F)$. First we need the following proposition. Let $\mathcal{T}$ be the collection of all sets $T \subseteq J_k$ which contain one edge of index $\mathbf{i}$ for each $\mathbf{i} \in I(F)$.

PROPOSITION 7.1. *The following statements are equivalent.*

(i) $\mathbf{1} \in PC(\{\chi(T) : T \in \mathcal{T}\})$.
(ii) $k|I(F)|\mathbf{1}/rn \in CH(\{\chi(T) : T \in \mathcal{T}\})$.
(iii) J_k *admits an F-balanced fractional perfect matching in which at most* $|I(F)|(rn + 1)$ *edges have non-zero weight.*
(iv) J_k *admits an F-balanced fractional perfect matching.*

PROOF. Let $X = \{\chi(T) : T \in \mathcal{T}\}$. Suppose first that (i) holds, so that $\mathbf{1} = \sum_{\mathbf{x} \in X} c_\mathbf{x} \mathbf{x}$ with $c_\mathbf{x} \geq 0$ for $\mathbf{x} \in X$. Then we have

$$k|I(F)| \sum_{\mathbf{x} \in X} c_\mathbf{x} = \sum_{\mathbf{x} \in X} c_\mathbf{x} \mathbf{x} \cdot \mathbf{1} = \mathbf{1} \cdot \mathbf{1} = rn.$$

Since any $\mathbf{x} \in X$ has non-negative integer coordinates we also have $c_\mathbf{x} \leq 1$ for each $\mathbf{x}$. Then by multiplying each $c_\mathbf{x}$ by $k|I(F)|/rn$ we obtain (ii). Now suppose that (ii) holds. Then by Theorem 3.1 we may write $k|I(F)|\mathbf{1}/rn = \sum_{\mathbf{x} \in X} c_\mathbf{x} \mathbf{x}$ with $c_\mathbf{x} \geq 0$ for $\mathbf{x} \in X$ so that at most $rn + 1$ of the $c_\mathbf{x}$'s are non-zero. For each $\mathbf{x} \in X$ assign weight $w_T = rnc_\mathbf{x}/k|I(F)|$ to some $T \in \mathcal{T}$ with $\chi(T) = \mathbf{x}$. Then assigning to each edge $e \in J_k$ the weight $w_e := \sum_{T \in \mathcal{T}:\ e \in T} w_T$ gives an F-balanced fractional perfect matching in which at most $|I(F)|(rn+1)$ edges have non-zero weight, so we have (iii). Trivially (iii) implies (iv), so it remains to show that (iv) implies (i).

Consider an F-balanced fractional perfect matching in J_k, where w_e denotes the weight of an edge e, so $\sum_{e \in J_k:\ \mathbf{i}(e) = \mathbf{i}} w_e$ is constant over $\mathbf{i} \in I(F)$. We assign weights to sets $T \in \mathcal{T}$ and modify the weights of edges $e \in J_k$ according to the following algorithm. Suppose at some step we have weights w'_e for $e \in J_k$, such that $\sum_{e \in J_k:\ \mathbf{i}(e)=\mathbf{i}} w'_e$ is constant over $\mathbf{i} \in I(F)$. Suppose that not all weights are zero, and choose $e_0 \in J_k$ with the smallest non-zero weight. For every $\mathbf{i} \in I(F)$ with $\mathbf{i} \neq \mathbf{i}(e_0)$, since

$$\sum_{e \in J_k:\ \mathbf{i}(e)=\mathbf{i}} w'_e = \sum_{e \in J_k:\ \mathbf{i}(e)=\mathbf{i}(e_0)} w'_e \neq 0,$$

we may choose $e_\mathbf{i}$ in J_k with $\mathbf{i}(e_\mathbf{i}) = \mathbf{i}$ and $w'_{e_\mathbf{i}} \neq 0$. Note that $w'_{e_\mathbf{i}} \geq w'_{e_0}$, by minimality of w'_{e_0}. Let $T \in \mathcal{T}$ consist of e_0 and the edges $e_\mathbf{i}$ for $\mathbf{i} \in I(F)$ with $\mathbf{i} \neq \mathbf{i}(e_0)$. We assign weight w'_{e_0} to T, and define new weights by $w''_{e_0} = 0$, $w''_{e_\mathbf{i}} = w'_{e_\mathbf{i}} - w'_{e_0}$, and $w''_e = w'_e$ otherwise. Then, with these new weights, $\sum_{e \in J_k:\ \mathbf{i}(e)=\mathbf{i}} w''_e$ remains constant over $\mathbf{i} \in I(F)$, and $w''_e \geq 0$ for every $e \in J_k$. Since the number of edges of zero weight has increased by at least one, after at most $|J_k|$ iterations every edge will have zero weight, at which point we stop. Then for any $\mathbf{x} \in X$, we let $c_\mathbf{x}$ be the sum of the weights assigned to any T with $\chi(T) = \mathbf{x}$. By construction we have $\mathbf{1} = \sum_{\mathbf{x} \in X} c_\mathbf{x} \mathbf{x}$, as required. □

The following lemma generalises Lemma 3.6 to the minimum F-degree setting. Indeed, Lemma 3.6 is the $\alpha = 0$ 'furthermore' statement of Lemma 7.2, applied with $r = 1$ and F generated by the unique function $f : [k] \to [1]$. The main statement in Lemma 7.2 shows that the same conclusion holds under a slightly weaker F-degree sequence if there is no space barrier.

LEMMA 7.2. *Suppose that $1/n \ll \alpha' \ll \beta, 1/D_F, 1/k, 1/r$ and that $\alpha \leq \alpha'$. Let V be a set partitioned into parts $V_1, \ldots, V_r$ each of size n, and F be a (k,r)-uniform connected allocation with $|F| \leq D_F$. Also let J be a $\mathcal{P}F$-partite k-system on V such that*

(i) $\delta^F(J) \geq \left(n, \frac{(k-1)n}{k} - \alpha n, \frac{(k-2)n}{k} - \alpha n, \ldots, \frac{n}{k} - \alpha n\right)$, *and*
(ii) *for any $p \in [k-1]$ and sets $S_i \subseteq V_i$ with $|S_i| = \lfloor pn/k \rfloor$ for $i \in [r]$ we have $|J_{p+1}[S]| \geq \beta n^{p+1}$, where $S := \bigcup_{i \in [r]} S_i$.*

Then J_k admits an F-balanced fractional perfect matching. Furthermore, if $\alpha = 0$ then this conclusion holds even without assuming condition (ii), any lower bound on n, or that F is connected.

PROOF. First we give a construction that reduces to the case when $k \mid n$. For $i \in [r]$, let V'_i be a set of size kn consisting of copies $v_i(j)$, $j \in [k]$ of each $v_i \in V_i$. Let J' be the k-system whose edges are all possible copies of the edges of J. Then $\delta^F(J') = k\delta^F(J)$, so (i) holds for J'. To see that (ii) holds also, fix $p \in [k-1]$ and sets $S_i' \subseteq V_i'$ with $|S_i'| = pn$ for $i \in [r]$. Let $S_i \subseteq V_i$ consist of all vertices $v_i \in V_i$ with a copy $v_i(j)$ in S_i'. Then $|S_i| \geq |S_i'|/k \geq \lfloor pn/k \rfloor$. So $|J_{p+1}[S]| \geq \beta n^{p+1}$ by (ii), where $S := \bigcup_{i \in [r]} S_i$. Each edge of $J_{p+1}[S]$ has at least one copy in $J'_{p+1}[S']$, where $S' := \bigcup_{i \in [r]} S_i'$. We deduce that $|J'_{p+1}[S']| \geq \beta n^{p+1} \geq (\beta/k^{p+1})(kn)^{p+1}$, so (ii) holds for J' with β/k^{p+1} in place of β. Assuming the result when $k \mid n$, we find that J_k' admits an F-balanced fractional perfect matching. From this we obtain an F-balanced fractional perfect matching in J_k, where the weight of an edge is obtained by combining the weights of its copies in J_k' and dividing by k. Thus we can assume $k \mid n$.

Now suppose for a contradiction that J_k has no F-balanced fractional perfect matching. Let $\mathcal{T}$ be as above; then Proposition 7.1 implies that $\mathbf{1} \notin PC(\{\chi(T) : T \in \mathcal{T}\})$. So by Farkas' Lemma (Lemma 3.4), there is some $\mathbf{a} \in \mathbb{R}^{rn}$ such that $\mathbf{a} \cdot \mathbf{1} < 0$ and $\mathbf{a} \cdot \chi(T) \geq 0$ for every $T \in \mathcal{T}$. Note that any F-balanced multiset E in J_k can be expressed as $\sum_{i=1}^{z} T_i$ for some $T_1, \ldots, T_z$ in $\mathcal{T}$, and so satisfies $\mathbf{a} \cdot \chi(E) \geq 0$. For $i \in [r]$ let $v_{i,1}, \ldots, v_{i,n}$ be the vertices of V_i, and let $a_{i,1}, \ldots, a_{i,n}$ be the corresponding coordinates of $\mathbf{a}$, where the vertex labels are chosen so that $a_{i,1} \leq a_{i,2} \leq \cdots \leq a_{i,n}$ for each $i \in [r]$.

For any multisets S and S' in V of equal size m, we say that S dominates S', and write $S' \leq S$, if we may write $S = \{v_{i_1,j_1}, \ldots, v_{i_m,j_m}\}$ and $S' = \{v_{i_1',j_1'}, \ldots, v_{i_m',j_m'}\}$ so that for each $\ell \in [m]$ we have $i_\ell = i_\ell'$ and $j_\ell' \leq j_\ell$. Note that $\leq$ is a transitive binary relation. We also observe that if $S' \leq S$ then $\mathbf{a} \cdot \chi(S') \leq \mathbf{a} \cdot \chi(S)$. As usual, for a multiset E in J_k we write $\chi(E) = \sum_{e \in E} \chi(e)$, thus identifying E with the multiset in V in which the multiplicity of $v \in V$ is the number of edges in E containing it, counting with repetition. We recall that $S + T$ denotes the multiset union of two multisets S and T, and for $i \in \mathbb{N}$, iS denotes the multiset union of i copies of S. We extend our 'arithmetic' of multisets to include subtraction, writing $S - T$ for the multiset A such that $A + T = S$, if it exists. It will also be convenient to manipulate formal expressions $S - T$ that do not correspond to multisets, via the rule $(S - T) + (S' - T') = (S + S') - (T + T')$, which can be understood as a shorthand for $\chi(S) - \chi(T) + \chi(S') - \chi(T') = \chi(S) + \chi(S') - \chi(T) + \chi(T')$.

We start by proving the $\alpha = 0$ statement. For a mental picture, it is helpful to think of the vertices arranged in a grid, with r columns corresponding to the parts V_i, $i \in [r]$, and k rows, where the subsquare in column i and row j contains the vertices $v_{i,(j-1)n/k+s}$, $s \in [n/k]$. We partition V into sets $(W_s : s \in [n/k])$, where

$$W_s = \{v_{i,(j-1)n/k+s} : i \in [r], j \in [k]\}$$

consists of all vertices at height s, for each column i and row j. Note that for each $s \in [n/k]$ we have $W_s \geq W_1$. Thus we have

$$0 > \mathbf{a} \cdot \mathbf{1} = \sum_{s \in [n/k]} \mathbf{a} \cdot \chi(W_s) \geq (n/k) \mathbf{a} \cdot \chi(W_1).$$

To obtain the required contradiction, we will show that a constant multiple of W_1 dominates an F-balanced multiset of edges of J_k. To see that this suffices, define

$$X^f := \{v_{f(j),(j-1)n/k+1} : j \in [k]\}$$

for $f \in F$. Since F is (k, r)-uniform we have $\sum_{f \in F} X^f = |F| r^{-1} W_1$. Now, by the minimum F-degree of J, we may greedily form an edge $e^f = \{v_{f(1),d_1}, \ldots, v_{f(k),d_k}\} \in J$ with $d_1 = 1$ and $d_j \leq (j-1)n/k + 1$ for each $2 \leq j \leq k$. Then X^f dominates

e^f, so $|F|r^{-1}W_1 = \sum_{f \in F} X^f$ dominates the F-balanced multiset $\{e^f : f \in F\}$. It follows that $\mathbf{a} \cdot \chi(W_1) \geq 0$, so we have the required contradiction to the assumption that J_k has no F-balanced fractional perfect matching.

Now consider the case $0 < \alpha < \alpha'$. We will obtain a contradiction by a similar strategy to that used when $\alpha = 0$, namely partitioning V into 'dominating' sets, where we say a multiset S is *dominating* if some constant multiple of S dominates an F-balanced multiset of edges in J_k. Note that if S is dominating we have $\mathbf{a} \cdot \chi(S) \geq 0$, so this will give the contradiction $\mathbf{a} \cdot \mathbf{1} \geq 0$. We may assume that $C := \alpha n$ is an integer. We also let N be an integer with $\alpha' \ll 1/N \ll 1/D_F, 1/k, 1/r$ and $|F| \mid N$. Now we define sets W_s, $s \in [n/k - CN]$ of size rk by

$$W_s := \{v_{i,s} : i \in [r]\} \cup \{v_{i,(j-1)n/k+C+s} : i \in [r], 2 \leq j \leq k\}.$$

Note that this agrees with our previous definition in the case when $C = \alpha = 0$; now we have increased by C the height of the vertices in rows 2 to k. Again we have $W_s \geq W_1$, and we will show that a constant multiple of W_1 dominates an F-balanced multiset of edges. To see this, define

$$X^f := \{v_{f(1),1}\} \cup \{v_{f(j),(j-1)n/k+C+1} : 2 \leq j \leq k\}$$

for $f \in F$ (again, this agrees with our previous notation when $C = \alpha = 0$). Since F is (k,r)-uniform we have $\sum_{f \in F} X^f = |F|r^{-1}W_1$. By the minimum F-degree of J, we may greedily form an edge $e^f = \{v_{f(1),d_1}, \ldots, v_{f(k),d_k}\} \in J$ with $d_1 = 1$ and $d_j \leq (j-1)n/k + C + 1$ for each $2 \leq j \leq k$. Then X^f dominates e^f, so $\sum_{f \in F} X^f = |F|r^{-1}W_1$ dominates $\{e^f : f \in F\}$.

Thus we have arranged that most vertices in V belong to dominating sets W_s, but in each column we still need to deal with the CN highest vertices in row 1, and the C lowest and $C(N-1)$ highest vertices in rows 2 to k. We will partition these into sets Z_s, $s \in [C]$ of size rkN, so that each Z_s contains N vertices in each of the rk subsquares of the grid of vertices, and in rows 2 to k these comprise 1 'low' vertex and $N - 1$ 'high' vertices. The formal definition is as follows. Define $P := [k] \times \{0, \ldots, N-1\} \cup \{(1,N)\} \setminus \{(k,0)\}$. Then for each $s \in [C]$, let

$$Z_s := \{v_{i,jn/k-tC+s} : (j,t) \in P, i \in [r]\}.$$

Note that the sets W_s, $s \in [n/k - CN]$ and Z_s, $s \in [C]$ partition V. Next we define a multiset that is dominated by each Z_s. Write $y_{i,j} = v_{i,jn/k-CN}$ for $i \in [r], j \in [k]$,

$$Y = \{y_{i,j} : i \in [r], j \in [k]\}, \text{ and } D = NY + \{y_{i,1} : i \in [r]\} - \{y_{i,k} : i \in [r]\}.$$

Thus the CN highest vertices in the subsquare of column i and row j are each above the corresponding vertex $y_{i,j}$ of Y. We claim that each Z_s dominates D. To see this, note that for each $i \in [r]$ and $2 \leq j \leq k-1$, the N copies of $y_{i,j}$ in D are dominated by the $N - 1$ high vertices $v_{i,jn/k-tC+s}$, $t \in [N-1]$ in row j and the low vertex $v_{i,jn/k+s}$ in row $j+1$. In row k we have removed one copy of each $y_{i,k}$ from D, so the remaining $N - 1$ copies are dominated by the high vertices in row k. In row 1 we have $N + 1$ copies of $y_{i,j}$ in D, which are dominated by the N vertices $v_{i,n/k-tC+s}$, $t \in [N]$ in row 1 and the low vertex $v_{i,n/k+s}$ in row 2. Thus $Z_s \geq D$.

The remainder of the proof is showing that D is a dominating multiset; we divide this into 3 claims. The first claim exploits the absence of a space barrier to find edges that are lower than those guaranteed by the minimum degree condition. Whereas the minimum degree condition gives edges in which the ith vertex is (close to being) in row i or below for $i \in [k]$, for each $p \in [k-1]$ we can find an edge where the first $p+1$ vertices are in row p or below, and the ith vertex is (close to being) in row i or below for $p+2 \leq i \leq k$. Intuitively, we can think of such an edge as having a 'p-demoted' vertex, in that the $(p+1)$st vertex is lower than guaranteed by the minimum degree condition (although the first p vertices may be higher). We also require that a demoted vertex is not too near the top of the row it has demoted too, in that it is below the corresponding vertex $y_{i,j}$. In the first claim we have no control over which of the parts V_i contains a demoted vertex, so in the second claim we exploit the connectivity of F to construct a multiset with a demoted vertex in any desired part. Finally, in the third claim we take an appropriate non-negative linear combination of multisets with p-demoted vertices for all p to obtain a multiset dominated by D.

For the first claim we need edges dominated by one of the sets

$$B_p^f := \{y_{f(j),p} : j \in [p+1]\} \cup \{y_{f(j),j} : p+2 \leq j \leq k\}.$$

CLAIM 7.3. *For any $p \in [k-1]$ there is some $f_p \in F$ and $e_p \in J_k$ with $e_p \leq B_p^{f_p}$.*

To prove the claim, for each $i \in [r]$ let $S_i = \{v_{i,d} : d \leq pn/k - CN\}$, and arbitrarily choose S_i' of size pn/k containing S_i. Let $S = \bigcup_{i \in [r]} S_i$ and $S' = \bigcup_{i \in [r]} S_i'$. Then by condition (ii) we have $|J_{p+1}[S']| \geq \beta n^{p+1}$. At most $rCN(rn)^p < \beta n^{p+1}$ edges of J_{p+1} intersect $S' \setminus S$, so $|J_{p+1}[S]| > 0$. We may therefore choose an edge $e = \{v_{f(1),d_1}, \ldots, v_{f(p+1),d_{p+1}}\} \in J_{p+1}$ where $d_j \leq pn/k - CN$ for $j \in [p+1]$, for some function $f : [p+1] \to [r]$. Since J is $\mathcal{P}F$-partite, f must be the restriction of some $f_p \in F$. Then by the minimum f_p-degree assumption on J we can greedily extend e to an edge $e_p = \{v_{f(1),d_1}, \ldots, v_{f(k),d_k}\}$ with $d_j \leq (j-1)n/k + C + 1$ for $p+2 \leq j \leq k$. Thus $B_p^{f_p}$ dominates e_p, which proves Claim 7.3.

For the second claim we will find F-balanced multisets of edges dominated by the multisets

$$D_p^\ell = 2k|F|r^{-1}Y + (p+1)\{y_{\ell,p}\} - \{y_{\ell,j} : j \in [p+1]\}.$$

CLAIM 7.4. *For any $p \in [k-1]$ and $\ell \in [r]$ there is an F-balanced multiset E_p^ℓ in J_k with $E_p^\ell \leq D_p^\ell$.*

To prove the claim, we start by applying Claim 7.3, obtaining $e_p \in J_k$ and $f_p \in F$ with $e_p \leq B_p^{f_p}$. Since F is connected, there is a connected graph G_F on $[r]$ such that for every $ii' \in E(G_F)$ and $j, j' \in [k]$ with $j \neq j'$ there is $f \in F$ with $f(j) = i$ and $f(j') = i'$. Choose for each $j \in [p+1]$ a path $f_p(j) = i_1^j, \ldots, i_{s_j+1}^j = \ell$ in G_F from $f_p(j)$ to ℓ. For each $z \in [s_j]$, let $f_z^j \in F$ be such that $f_z^j(j) = i_z^j$ and $f_z^j(p) = i_{z+1}^j$, and let $\hat{f}_z^j$ be obtained from f_z^j by swapping the values of $f_z^j(j)$ and $f_z^j(p)$. Since F is invariant under permutation we have $\hat{f}_z^j \in F$. Now recall that for each $f \in F$ we have an edge $e^f \leq X^f$, where $X^f = \{v_{f(j),(j-1)n/k+C+1} : j \in [k]\}$. We define $Y^f := \{y_{f(j),j} : j \in [k]\}$, and note that $Y^f \geq X^f \geq e^f$ and $\sum_{f \in F} Y^f = |F|r^{-1}Y$. Next we show that we can write D_p^ℓ as

(2) $$D_p^\ell = 2k \sum_{f \in F} Y^f + (B_p^{f_p} - Y^{f_p}) + \sum_{j \in [p+1]} \sum_{z \in [s_j]} (Y^{f_z^j} - Y^{\hat{f}_z^j}).$$

To see this, note that $Y^{f_z^j} - Y^{\hat{f}_z^j} = \{y_{f_z^j(j),j}, y_{f_z^j(p),p}\} - \{y_{f_z^j(j),p}, y_{f_z^j(p),j}\} = (\{y_{i_z^j,j}\} - \{y_{i_z^j,p}\}) - (\{y_{i_{z+1}^j,j}\} - \{y_{i_{z+1}^j,p}\})$ for each $j \in [p+1], z \in [s_j]$, so

$$\sum_{z \in [s_j]} (Y^{f_z^j} - Y^{\hat{f}_z^j}) = (\{y_{f_p(j),j}\} - \{y_{f_p(j),p}\}) - (\{y_{\ell,j}\} - \{y_{\ell,p}\})$$

for each $j \in [p+1]$. Then

$$\sum_{j \in [p+1]} \sum_{z \in [s_j]} (Y^{f_z^j} - Y^{\hat{f}_z^j}) = \sum_{j \in [p+1]} (\{y_{f_p(j),j}\} - \{y_{f_p(j),p}\}) - \sum_{j \in [p+1]} (\{y_{\ell,j}\} - \{y_{\ell,p}\})$$
$$= Y^{f_p} - B_p^{f_p} + (p+1)\{y_{\ell,p}\} - \{y_{\ell,j} : j \in [p+1]\},$$

which proves (2). To define E_p^ℓ, we take $2k$ copies of e^f for each $f \in F$, replace one copy of e^{f_p} by a copy of e_p, and replace one copy of $e^{\hat{f}_z^j}$ by one copy of $e^{f_z^j}$ for each $j \in [p+1], z \in [s_j]$; note that there are enough copies for these replacements, as $f_z^j, z \in [s_j]$ are distinct. Thus E_p^ℓ contains $2k|F|$ edges and is F-balanced, since each edge was replaced by another of the same index. To see that $E_p^\ell \leq D_p^\ell$, we assign edges in E_p^ℓ to terms on the right hand side of (2), counted with multiplicity according to their coefficient, such that each edge of E_p^ℓ is dominated by its assigned term. Before the replacements, we have $2k$ copies of e^f for each $f \in F$, which we assign to the $2k$ copies of Y^f. To account for the replacement of a copy of e^{f_p} by e_p, we remove one of the assignments to Y_{f_p} and assign the replaced copy of e_p to $B_p^{f_p}$. To replace a copy of $e^{\hat{f}_z^j}$ by $e^{f_z^j}$, we remove one of the assignments to $Y^{\hat{f}_z^j}$ and assign the replaced copy of $e^{f_z^j}$ to $Y^{f_z^j}$. Thus we have an assignment showing that $E_p^\ell \leq D_p^\ell$, which proves Claim 7.4.

Finally, we will establish the required property of D by taking an appropriate non-negative linear combination of the multisets E_p^ℓ and D_p^ℓ. This is given by the following claim.

CLAIM 7.5. *There are non-negative integer coefficients m_j, $j \in [k]$ such that*

$$M := \sum_{j=1}^{k} m_j \leq k^k \text{ and } D = (N - 2k|F|M)Y + \sum_{\ell \in [r]} \sum_{p \in [k-1]} m_p D_p^\ell.$$

To prove the claim, we define the coefficients recursively by $m_k = 0$, $m_{k-1} = 1$, $m_{k-2} = k-1$ and

$$m_j = (j+2)m_{j+1} - (j+3)m_{j+2} \text{ for } 1 \leq j \leq k-3.$$

Since $m_j \leq km_{j+1}$ for any $j \in [k-1]$ we have $M \leq k^k$. Next we show that

$$(3) \qquad (s+1)m_s - \sum_{j=s-1}^{k} m_j = \begin{cases} -1 & s = k, \\ 0 & k-1 \geq s \geq 2, \\ 1 & s = 1. \end{cases}$$

For $s = k$ we have $(k+1)m_k - m_{k-1} - m_k = -1$, and for $s = k-1$ we have $km_{k-1} - \sum_{j=k-2}^{k} m_j = k - (k-1) - 1 = 0$. Also, for $k-2 \geq s \geq 2$ we have

$$(s+1)m_s - \sum_{j=s-1}^{k} m_j = m_{s-1} + (s+2)m_{s+1} - \sum_{j=s-1}^{k} m_j = (s+2)m_{s+1} - \sum_{j=s}^{k} m_j,$$

so these cases follow from the case $s = k - 1$. Finally, writing $m_0 = 0$, we have

$$\sum_{s=1}^{k} \left((s+1)m_s - \sum_{j=s-1}^{k} m_j \right) = m_k - m_0 = 0;$$

this implies the case $s = 1$, and so finishes the proof of (3). Now we show by induction that $m_j \geq (j+2)m_{j+1}/2$ for $j = k-2, k-3, \ldots, 2$. For the base case $j = k-2$ we have $m_{k-2} = k \geq k/2 = km_{k-1}/2$. Now suppose $2 \leq j \leq k-3$ and $m_{j+1} \geq (j+3)m_{j+2}/2$. Then $m_j - (j+2)m_{j+1}/2 \geq (j+2)m_{j+1}/2 - (j+3)m_{j+2} \geq (j-2)m_{j+1}/2 \geq 0$. Therefore $m_j \geq 0$ for $2 \leq j \leq k$. Then by (3), we also have $m_1 = (1 + \sum_{j=0}^{k} m_j) \geq 0$. Now by definition of D_p^ℓ, for any $\ell \in [r]$ we have

$$\sum_{p \in [k-1]} m_p D_p^\ell = \sum_{p \in [k-1]} m_p(2k|F|r^{-1}Y + (p+1)\{y_{\ell,p}\} - \{y_{\ell,j} : j \in [p+1]\})$$

$$= 2Mk|F|r^{-1}Y + \sum_{s \in [k]} \left((s+1)m_s - \sum_{j=s-1}^{k} m_j \right) \{y_{\ell,s}\}.$$

Summing over ℓ and applying (3), we obtain $\sum_{\ell \in [r]} \sum_{p \in [k-1]} m_p D_p^\ell = 2Mk|F|Y + \{y_{\ell,1} : \ell \in [r]\} - \{y_{\ell,k} : \ell \in [r]\}$. By definition of D, this proves Claim 7.5.

To finish the proof of the lemma, we show that D is dominating, by defining an F-balanced multiset E in J_k with $E \leq D$. We let E be the combination of $Nr/|F| - 2Mrk$ copies of $\{e^f : f \in F\}$ with m_p copies of E_p^ℓ for each $\ell \in [r]$ and $p \in [k]$ (note that $Nr/|F| - 2Mrk$ is positive since $N \gg D_F, k, r$). Then E is F-balanced, since each E_p^ℓ is F-balanced. Now recall from the proof of Claim 7.4 that we can write $\sum_{f \in F} Y^f = |F|r^{-1}Y$ with $Y^f \geq e^f$ for $f \in F$. Substituting this in the expression for D in Claim 7.5 we can see that D dominates E termwise: each Y^f in D dominates an e^f in E, and each D_p^ℓ in D dominates an E_p^ℓ in E. Now recall that V is partitioned into sets W_s, $s \in [n/k - CN]$ and Z_s, $s \in [C]$, where each W_s dominates the dominating set W_1, and each Z_s dominates D, which we have now shown is dominating. We deduce that $\mathbf{a} \cdot \mathbf{1} \geq 0$, which gives the required contradiction to the assumption that J_k has no F-balanced fractional perfect matching. □

7.2. Almost perfect matchings

In this section we prove a lemma that will be used in the following section to prove a common generalisation of Theorems 2.4 and 2.7, in which we find a matching covering all but a constant number of vertices, and a common generalisation of Theorems 2.9 and 2.10, where we find a perfect matching under the additional assumption that there is no divisibility barrier. The following lemma is a weaker version of the common generalisation of Theorems 2.4 and 2.7. It states that if a k-system J satisfies our minimum F-degree condition and does not have a space barrier, then J_k contains a matching which covers all but a small proportion of $V(J)$. One should note that the proportion ψ of uncovered vertices can be made much smaller than the deficiency α in the F-degree sequence. The proof will also show that a slightly stronger degree condition yields the same conclusion even in the presence of a space barrier, from which we shall deduce a common generalisation of Theorems 2.2 and 2.5 on almost perfect matchings.

LEMMA 7.6. *Suppose that $1/n \ll \psi \ll \alpha \ll \beta, 1/D_F, 1/r, 1/k$. Let $\mathcal{P}$ be a partition of a set V into parts $V_1, \ldots, V_r$ of size n and F be a connected (k,r)-uniform allocation with $|F| \leq D_F$. Suppose that J is a $\mathcal{P}F$-partite k-system on V such that*

(i) $\delta^F(J) \geq \left(n, \left(\frac{k-1}{k} - \alpha\right) n, \left(\frac{k-2}{k} - \alpha\right) n, \ldots, \left(\frac{1}{k} - \alpha\right) n\right)$, *and*
(ii) *for any $p \in [k-1]$ and sets $S_i \subseteq V_i$ with $|S_i| = \lfloor pn/k \rfloor$ for $i \in [r]$ we have $|J_{p+1}[S]| \geq \beta n^{p+1}$, where $S := \bigcup_{i \in [r]} S_i$.*

Then J_k contains an F-balanced matching M which covers all but at most ψn vertices of J.

PROOF. Introduce new constants with

$$1/n \ll \varepsilon \ll d_a \ll 1/a \ll \nu, 1/h \ll c_k \ll \cdots \ll c_1 \ll \gamma \ll \psi \ll \alpha \ll \beta, 1/D_F, 1/r, 1/k.$$

We may additionally assume that $r \mid h$. Since $r \mid |V(J)|$ and $r \mid a!h$, we may delete up to $a!h$ vertices of J so that equally many vertices are deleted from each vertex class and the number of vertices remaining in each part is divisible by $a!h$. By adjusting the constants, we can assume for simplicity that in fact $a!h$ divides n, so no vertices were deleted. Fix any partition $\mathcal{Q}$ of V into h parts of equal size which refines $\mathcal{P}$, and let J' be the k-system on V formed by all edges of J which are $\mathcal{Q}$-partite. Then J'_k is a $\mathcal{Q}$-partite k-graph on V, so by Theorem 6.2 there is an a-bounded vertex-equitable $\mathcal{Q}$-partition $(k-1)$-complex P on V and a $\mathcal{Q}$-partite k-graph G on V that is ν/r^k-close to J'_k and perfectly ε-regular with respect to P. Let $Z := G \triangle J'_k$. Then since G is ν/r^k-close to J'_k we have $|Z| \leq \nu n^k$. Also let $W_1, \ldots, W_{rm_1}$ be the clusters of P (note that $rm_1 \leq ah$), and let n_1 be their common size, so $n_1 m_1 = n$.

Consider the reduced k-system $R := R^{J'Z}_{\mathcal{P}\mathcal{Q}}(\nu, \mathbf{c})$. Recall that R has vertex set $[rm_1]$, where vertex i corresponds to cluster W_i of P, and that $\mathcal{P}_R$ and $\mathcal{Q}_R$ are the partitions of $[rm_1]$ corresponding to $\mathcal{P}$ and $\mathcal{Q}$ respectively. So R is $\mathcal{P}_R F$-partite and $\mathcal{Q}$-partite. For each $i \in [r]$ let U_i be the part of $\mathcal{P}_R$ corresponding to part V_i of $\mathcal{P}$; since P was vertex-equitable each part U_i has size m_1. We now show that since J had no space barrier, R also does not have a space barrier, even after the deletion of a small number of vertices from each U_i.

CLAIM 7.7. *Suppose that sets $U'_i \subseteq U_i$ for $i \in [r]$ satisfy $|U'_1| = \cdots = |U'_r| = m' \geq (1-\alpha)m_1$. Let $U' := \bigcup_{i \in [r]} U'_i$, $R' := R[U']$ and $\mathcal{P}_{R'}$ be the restriction of $\mathcal{P}_R$ to U'. Then for any $p \in [k-1]$ and sets $S'_i \subseteq U'_i$ with $|S'_i| = \lfloor pm'/k \rfloor$ for $i \in [r]$ we have $|R'_{p+1}[S']| \geq \beta(m')^{p+1}/10$, where $S' := \bigcup_{i \in [r]} S'_i$.*

To prove the claim, let $S = \bigcup_{i \in [r]} S_i$, where $S_i = \bigcup_{j \in S'_i} W_j \subseteq V_i \subseteq V$ for $i \in [r]$. Then $|S'_i| \geq n_1 \lfloor pm'/k \rfloor \geq (1-2\alpha)pn/k$ for $i \in [r]$. Let $S'' = \bigcup_{i \in [r]} S''_i$, where for $i \in [r]$ we take S''_i to be any set of size $\lfloor pn/k \rfloor$ such that $S_i \subseteq S''_i$ if $|S_i| \leq \lfloor pn/k \rfloor$ and $S''_i \subseteq S_i$ if $|S_i| \geq \lfloor pn/k \rfloor$. By assumption (ii) on J we have $|J_{p+1}[S'']| \geq \beta n^{p+1}$, and so $|J'_{p+1}[S'']| \geq \beta n^{p+1}/2$, since at most n^{p+1}/h edges of J_{p+1} are not edges of J'_{p+1}. Since at most $2\alpha r n$ vertices of S'' are not vertices of S, it follows that $|J'_{p+1}[S]| \geq \beta n^{p+1}/4$. Letting $\mathcal{S}$ denote the partition of $V(J')$ into parts S and $V(J') \setminus S$, we can rephrase this as at least $\beta n^{p+1}/4$ edges $e \in J'_{p+1}$ have $\mathbf{i}_\mathcal{S}(e) = (p+1, 0)$. By Lemma 6.10 it follows that at least $\beta m_1^{p+1}/8$ edges $e \in R_{p+1}$ have $\mathbf{i}_{\mathcal{S}_R}(e) = (p+1, 0)$. Since $\mathcal{S}_R$ is the partition of $[m_1]$ into S' and $[m_1] \setminus S'$, we conclude that $|R_{p+1}[S']| \geq \beta m_1^{p+1}/8$, and therefore $|R'_{p+1}[S']| \geq \beta(m')^{p+1}/10$, proving the claim.

Now, since any edge $e \in J'$ has $d_{J'}^F(e) \geq d_J^F(e) - \alpha n/2$, by Lemma 6.8 we have

$$\delta^F(R) \geq ((1 - k\nu^{1/2})m_1, (\delta_1(J)/n - \alpha/2)m_1, \ldots, (\delta_{k-1}(J)/n - \alpha/2)m_1)$$

with respect to $\mathcal{P}_R$. So there are at most $k\nu^{1/2}m_1$ vertices i in each part of $\mathcal{P}_R$ for which $\{i\}$ is not an edge of R. Thus we can delete $k\nu^{1/2}m_1$ vertices from each part to obtain R' with $m' = (1 - k\nu^{1/2})m_1 \geq (1-\alpha)m_1$ vertices in each part and

$$\delta^F(R') \geq (m', (\delta_1(J)/n - \alpha)m', \ldots, (\delta_{k-1}(J)/n - \alpha)m'). \quad (4)$$

Claim 7.7 and (4) together show that R'_k satisfies the conditions of Lemma 7.2, and so admits an F-balanced fractional perfect matching. By Proposition 7.1 it follows that R'_k admits a F-balanced fractional perfect matching in which there are at most $|I(F)|(rm' + 1)$ edges $e \in R'_k$ of non-zero weight. For $e \in R'_k$ let w_e be the weight of e in such a fractional matching. So $\sum_{e \ni v} w_e = 1$ for any $v \in V(R')$ and $\sum_{e \in R'_k:\ \mathbf{i}(e) = \mathbf{i}} w_e$ has common value $m'r/k|I(F)|$ for every $\mathbf{i} \in I(F)$. Next, partition each cluster W_i into parts $\{W_i^e : e \in R'_k\}$ such that

$$|W_i^e| = \begin{cases} w_e n_1 & \text{if } e \text{ is incident to vertex } i \text{ of } R', \\ 0 & \text{otherwise.} \end{cases}$$

The lemma will now follow easily from the following claim.

CLAIM 7.8. *For any $e \in R'_k$ there exists a matching M_e in $J_k[\bigcup_{i \in e} W_i^e]$ of size at least $(w_e - \gamma)n_1$.*

To prove the claim, first note that if $w_e \leq \gamma$ then there is nothing to prove, so we may assume that $w_e > \gamma$. Let M be a maximal matching in $J_k[\bigcup_{i \in e} W_i^e]$, and suppose for a contradiction that $|M| < (w_e - \gamma)n_1$. For each $i \in e$ let W'_i consist of the vertices in W_i^e not covered by M; then $|W'_i| \geq \gamma n_1$ for each $i \in e$. Now observe that since $e \in R'_k$, and therefore $e \in R_k$, we know that e is $\mathcal{Q}_R$-partite, that $|Z[\bigcup_{i \in e} W_i^e]| \leq \nu^{2^{-k}} n_1^k$, and that $|J'_k[\bigcup_{i \in e} W_i^e]| \geq c_k n_1^k$. Since $J'_k \setminus Z \subseteq G_k$ we therefore have $|G[\bigcup_{i \in e} W_i^e]| \geq c_k n_1^k/2$. Since G is perfectly ε-regular with respect to P, by Proposition 6.9 there is a k-partite k-complex G' whose vertex classes are W_i^e for $i \in e$ such that G' is ε-regular, $d_{[k]}(G') \geq c_k/4$, $d(G') \geq d_a$, $G'_k \subseteq G$, and $|Z \cap G'_k| \leq \nu^{2^{-k}/3}|G'_k|$. Writing $W' := \bigcup_{i \in e} W'_i$, it follows by Lemma 6.1 that $G'[W']$ has $d(G'[W']) \geq d(G')/2$. So $G'_k[W']$ contains at least $\gamma^k |G'_k|/2$ edges. Since $\nu \ll \gamma$, some edge of $G'_k[W']$ is not an edge of Z, and is therefore an edge of $J_k[W']$, contradicting the maximality of M. This proves Claim 7.8.

To finish the proof of Lemma 7.6, we apply Claim 7.8 to find matchings M_e for each edge $e \in R'_k$ with $w_e > 0$. Then the union $M := \bigcup_{e \in R'_k:\ w_e > 0} M^e$ of all these matchings is a matching in J_k. Furthermore, by choice of the weights w_e, for any $\mathbf{i} \in I(F)$ the number of edges $e' \in M$ with $\mathbf{i}(e') = \mathbf{i}$ is at least

$$\sum_{e \in R'_k: w_e > 0,\ \mathbf{i}(e) = \mathbf{i}} (w_e - \gamma)n_1 \geq \frac{rm'n_1}{k|I(F)|} - \gamma|I(F)|(rm'+1)n_1$$

$$\geq \frac{(1 - k\nu^{1/2})rm_1 n_1 - 2\gamma k D_F^2 rm_1 n_1}{k|I(F)|} \geq \frac{|V(J)| - \psi n}{k|I(F)|}.$$

So for each $\mathbf{i} \in I(F)$ we may choose $(|V(J)| - \psi n)/k|I(F)|$ edges $e' \in M$ with $\mathbf{i}(e') = \mathbf{i}$; these edges together form an F-balanced matching in J_k which covers all but at most ψn vertices of J. □

Examining the above proof, we note that condition (ii) and the connectedness of F were only used in the proof of Claim 7.7, which in turn was only used to show that R'_k admits an F-balanced fractional perfect matching. If we instead assume the stronger F-degree condition $\delta^F(J) \geq \left(n, \left(\frac{k-1}{k} + \alpha\right)n, \left(\frac{k-2}{k} + \alpha\right)n, \ldots, \left(\frac{1}{k} + \alpha\right)n\right)$ then we have $\delta^F(R') \geq \left(m', \frac{(k-1)m'}{k}, \ldots, \frac{m'}{k}\right)$ by (4). Then the 'furthermore' statement of Lemma 7.2 implies that R'_k admits an F-balanced fractional perfect matching. Thus we have also proved the following lemma.

LEMMA 7.9. *Suppose that $1/n \ll \psi \ll \alpha \ll 1/D_F, 1/r, 1/k$. Let $\mathcal{P}$ be a partition of a set V into parts $V_1, \ldots, V_r$ of size n and F be a (k,r)-uniform allocation with $|F| \leq D_F$. Suppose that J is a $\mathcal{P}F$-partite k-system on V such that*

$$\delta^F(J) \geq \left(n, \left(\frac{k-1}{k} + \alpha\right)n, \left(\frac{k-2}{k} + \alpha\right)n, \ldots, \left(\frac{1}{k} + \alpha\right)n\right).$$

Then J_k contains an F-balanced matching M which covers all but at most ψn vertices of J.

Now we deduce the common generalisation of Theorems 2.2 and 2.5. To do this, we add some 'fake' edges to increase the minimum degree of J, so that we may use Lemma 7.9. This gives an almost perfect matching in the new system. The fake edges are chosen so that only a small number of them can appear in any matching, so we then remove them to give an almost perfect matching in the original system.

THEOREM 7.10. *Suppose that $1/n \ll \alpha \ll 1/D_F, 1/r, 1/k$. Let $\mathcal{P}$ be a partition of a set V into sets $V_1, \ldots, V_r$ each of size n and F be a (k,r)-uniform allocation with $|F| \leq D_F$. Suppose that J is a $\mathcal{P}F$-partite k-system on V with*

$$\delta^F(J) \geq \left(n, \left(\frac{k-1}{k} - \alpha\right)n, \left(\frac{k-2}{k} - \alpha\right)n, \ldots, \left(\frac{1}{k} - \alpha\right)n\right).$$

Then J contains a matching covers all but at most $9k^2 r\alpha n$ vertices of J.

PROOF. Choose a set X consisting of $8k\alpha n$ vertices in each part V_i for $i \in [r]$ uniformly at random. We form a k-system J' on $V(J)$ whose edge set consists of every edge of J, and *fake edges*, which are every $S \in \binom{V(J)}{\leq k}$ which is $\mathcal{P}F$-partite and intersects X. We claim that with high probability we have

(5) $\quad \delta^F(J') \geq \left(n, \left(\frac{k-1}{k} + \alpha\right)n, \left(\frac{k-2}{k} + \alpha\right)n, \ldots, \left(\frac{1}{k} + \alpha\right)n\right).$

To see this, first observe that $\delta_0^F(J') = \delta_0^F(J) = n$. Also, if e is a fake edge, then $d_{J'}^F(e) \geq n - k$. Now consider any $f \in F$, $j \in [k-1]$ and $e \in J_j$ such that we may write $e = \{v_1, \ldots, v_j\}$ with $v_i \in V_{f(i)}$ for $i \in [j]$. Then there are at least $\left(\frac{k-j}{k} - \alpha\right)n$ vertices $v_{j+1} \in V_{f(j+1)}$ such that $\{v_1, \ldots, v_{j+1}\} \in J$. We can assume that there is a set S of $n/2k$ vertices $v_{j+1} \in V_{f(j+1)}$ such that $\{v_1, \ldots, v_{j+1}\} \notin J$ (otherwise we are done). Then $Y = |S \cap X|$ is hypergeometric with mean $4\alpha n$, so by Lemma 6.13 we have $Y \geq 2\alpha n$ with probability at least $1 - 2e^{-\alpha n/3}$. On this event, there are at least $(k-j)n/k + \alpha n$ vertices $v_{j+1} \in V_{f(j+1)}$ such that

$\{v_1, \ldots, v_{j+1}\} \in J'$. Taking a union bound over at most $|F|kn^{k-1}$ choices of f, j and e we see that (5) holds with high probability. Fix a choice of X such that (5) holds. Then by Lemma 7.9 J' contains a matching M' which covers all but at most αn vertices of J'. Since every fake edge intersects X, there can be at most $|X| = 8kr\alpha n$ fake edges in M; deleting these edges we obtain a matching M which covers all but at most $\alpha n + 8k^2 r\alpha n \leq 9k^2 r\alpha n$ vertices of J. □

To deduce Theorem 2.2 we apply Theorem 7.10 with $r = 1$ and F consisting of ($k!$ copies of) the unique function $f : [k] \to [1]$, so $\delta^F(J) = \delta^f(J) = \delta(J)$. Similarly, to deduce Theorem 2.5 we apply Theorem 7.10 with F consisting of all injections $f : [k] \to [r]$, so $\delta^F(J) = \delta^*(J)$ is the partite minimum degree sequence. Note that in the latter case an F-balanced matching is exactly our notion of a balanced matching from earlier.

7.3. Perfect matchings

In this section we prove a common generalisation of Theorems 2.9 and 2.10 on perfect matchings. Essentially the same proof will also give a common generalisation of Theorems 2.4 and 2.7. Before giving the proof, for the purpose of exposition we sketch an alternative argument for the case of graphs ($k = 2$). For simplicity we just consider the cases $r = 2$ and $r = 1$. Suppose first that $r = 2$, i.e. we have a bipartite graph G with parts V_1 and V_2 of size n, with $\delta(G) \geq (1/2 - \alpha)n$. Suppose that G does not have a perfect matching. Then by Hall's theorem, there is $S_1' \subseteq V_1$ with $|N(S_1')| < |S_1'|$. By the minimum degree condition, each of S_1' and $N(S_1')$ must have size $(1/2 \pm \alpha)n$. Let S_1 be a set of size $\lfloor n/2 \rfloor$ that either contains or is contained in S_1' and let S_2 be a set of size $\lfloor n/2 \rfloor$ that either contains or is contained in $V_2 \setminus N(S_1')$. Then $S_1 \cup S_2$ contains at most $2\alpha n^2$ edges, so we have a space barrier.

Now suppose that $r = 1$, i.e. we have a graph G on n vertices with $\delta(G) \geq (1/2 - \alpha)n$, where n is even. Suppose that G does not have a perfect matching. Then by Tutte's Theorem, there is a set $U \subseteq V(G)$ so that $G \setminus U$ has more than $|U|$ odd components (i.e. connected components with an odd number of vertices). Suppose first that $|U| < (1/2 - 2\alpha)n$. Then $\delta(G \setminus U) \geq \alpha n$, so $G \setminus U$ has at most α^{-1} components. It follows that $|U| < \alpha^{-1}$. Then $\delta(G \setminus U) \geq (1/2 - 2\alpha)n$, so $G \setminus U$ has at most 2 components. It follows that $|U|$ is 0 or 1. Since n is even, $|U| = 0$, and G has two odd components, i.e. a divisibility barrier. Now suppose that $|U| > (1/2 - 2\alpha)n$. Then all but at most $2\alpha n$ of the odd components of $G \setminus U$ are isolated vertices, i.e. G contains an independent set I of size $(1/2 - 4\alpha)n$. Let S be a set of size $n/2$ containing I. Then S contains at most $2\alpha n^2$ edges, so we have a space barrier.

THEOREM 7.11. *Let* $1/n \ll \gamma \ll \alpha \ll \beta, \mu \ll 1/D_F, 1/r, 1/k$. *Suppose F is a (k, r)-uniform connected allocation with $|F| \leq D_F$, and that $b = k|I(F)|/r$ divides n. Let $\mathcal{P}$ be a partition of a set V into parts $V_1, \ldots, V_r$ of size n, and J be a $\mathcal{P}F$-partite k-complex on V such that*

  (i) $\delta^F(J) \geq \left(n, \left(\frac{k-1}{k} - \alpha\right)n, \left(\frac{k-2}{k} - \alpha\right)n, \ldots, \left(\frac{1}{k} - \alpha\right)n\right)$,
  (ii) *for any $p \in \lfloor k - 1 \rfloor$ and sets $S_i \subseteq V_i$ such that $|S_i| = \lfloor pn/k \rfloor$ for each $i \in [r]$ there are at least βn^k edges of J_k with more than p vertices in $S := \bigcup_{i \in [r]} S_i$, and*

(iii) $L_{\mathcal{P}'}^{\mu}(J_k)$ *is complete with respect to $\mathcal{P}$ for any partition $\mathcal{P}'$ of $V(J)$ which refines $\mathcal{P}$ and whose parts each have size at least $n/k - \mu n$.*

Then J_k contains a perfect matching which γ-represents F.

PROOF. We follow the strategy outlined in Section 2.6. The first step is to use hypergraph regularity to decompose J into an exceptional set and some clusters with a matched reduced k-system. Introduce new constants with

$$1/n \ll \varepsilon \ll d^* \ll d_a \ll 1/a \ll \nu, 1/h \ll \theta \ll d, c \ll c'_k \ll \cdots \ll c'_1$$
$$\ll c_k \ll \cdots \ll c_1 \ll \psi \ll \gamma \ll 1/C \ll \alpha \ll \alpha' \ll \mu, \beta \ll 1/D_F, 1/r, 1/k.$$

We also assume that $r \mid h$. By uniformity of F, any matching in J_k which contains one edge of index $\mathbf{i}$ for each $\mathbf{i} \in I(F)$ covers b vertices in each part of $\mathcal{P}'$. Since $b \mid a!h$ and $b \mid n$, we can arbitrarily delete at most $a!h/b$ such matchings to make the number of vertices remaining in each part divisible by $a!h$. By adjusting the constants, we can assume for simplicity that in fact $a!h$ divides n, so no vertices were deleted. Fix any balanced partition $\mathcal{Q}$ of V into h parts which refines $\mathcal{P}$, and let J' be the subcomplex obtained by deleting from J all those edges which are not $\mathcal{Q}$-partite. Then J'_k is a $\mathcal{Q}$-partite k-graph with order divisible by $a!h$, so by Theorem 6.2 there exists an a-bounded ε-regular vertex-equitable $\mathcal{Q}$-partition $(k-1)$-complex P on $V(J)$ and a $\mathcal{Q}$-partite k-graph G on V that is ν/r^k-close to J'_k and perfectly ε-regular with respect to P. Let $Z = G \triangle J'_k$, so $|Z| \le \nu n^k$ and any edge of $G \setminus Z$ is also an edge of J_k. Note that since $r \mid h$ the number of clusters of P is divisible by r; let $W_1, \ldots, W_{rm_1}$ be the clusters of P. Since P is a-bounded we must have $rm_1 \le ah$. In addition, since P is vertex-equitable, each cluster W_i has the same size; let $n_1 := |W_1| = \cdots = |W_{rm_1}| = n/m_1$ be this common size.

Let $R^1 := R_{P\mathcal{Q}}^{J'Z}(\nu, \mathbf{c})$ and $R^2 := R_{P\mathcal{Q}}^{J'Z}(\nu, \mathbf{c}')$ be reduced k-systems. So R^1 and R^2 have the common vertex set $[rm_1]$ partitioned into r parts $U_1^1, \ldots, U_r^1$ by $\mathcal{P}_{R^1} = \mathcal{P}_{R^2}$, where for each $i \in [r]$ part U_i^1 of $\mathcal{P}_{R^1}$ corresponds to part V_i of $\mathcal{P}$. Note that since P is vertex-equitable each U_i^1 has size m_1. Also note that since R^2 has weaker density parameters than R^1, any edge of R^1 is also an edge of R^2. Now, since J is a k-complex, condition (ii) implies that for any $p \in [k-1]$ and sets $S_i \subseteq V_i$ with $|S_i| = \lfloor pn/k \rfloor$ for $i \in [r]$ we have $|J_{p+1}[S]| \ge \beta n^{p+1}$, where $S := \bigcup_{i \in [r]} S_i$. So the conditions of Theorem 7.6 hold. Since R^1 is defined here exactly as R was in the proof of Theorem 7.6, we have the following claim, whose proof is identical to that of Claim 7.7.

CLAIM 7.12. *Suppose that sets $U_i \subseteq U_i^1$ for $i \in [r]$ satisfy $|U_1| = \cdots = |U_r| = m \ge (1 - 2k\alpha)m_1$. Let $U := \bigcup_{i \in [r]} U_i$, $R := R^1[U]$ and let $\mathcal{P}_R$ be the restriction of $\mathcal{P}_{R^1}$ to U. Then for any $p \in [k-1]$ and sets $S_i \subseteq U_i$ with $|S_i| = \lfloor pm/k \rfloor$ for $i \in [r]$ we have $|R_{p+1}[S]| \ge \beta m^{p+1}/10$, where $S := \bigcup_{i \in [r]} S_i$.*

Since any edge $e \in J'$ has $d_{J'}^F(e) \ge d_J^F(e) - \alpha n/2$, by Lemma 6.8 we have

(6) $\quad \delta^F(R^1), \delta^F(R^2) \ge \left((1 - k\nu^{1/2})m_1, \left(\frac{k-1}{k} - 2\alpha\right)m_1, \ldots, \left(\frac{1}{k} - 2\alpha\right)m_1 \right).$

So there are at most $k\nu^{1/2}m_1$ vertices i in each part of $\mathcal{P}_{R^1}$ for which $\{i\}$ is not an edge of R^1. We can therefore delete $k\nu^{1/2}m_1$ vertices from each part to obtain

$R^0 \subseteq R^1$ with $m_0 = (1 - k\nu^{1/2})m_1$ vertices in each part and

(7) $$\delta^F(R^0) \geq \left(m_0, \left(\frac{k-1}{k} - 3\alpha\right) m_0, \ldots, \left(\frac{1}{k} - 3\alpha\right) m_0\right).$$

By (7) and Claim 7.12, R^0 satisfies the conditions of Lemma 7.6, so contains an F-balanced matching M which covers all but at most ψm_0 vertices of R^0. For each $i \in [r]$ let U_i be the vertices in the ith part of $\mathcal{P}_{R^1}$ covered by M; by uniformity of F each U_i has size $m \geq (1 - \psi/r)m_0 \geq (1 - 2\psi/r)m_1$. Let $\mathcal{P}_R$ be the restriction of $\mathcal{P}_{R^1}$ to $U := \bigcup_{j \in [r]} U_j$, so that the parts of $\mathcal{P}_R$ are U_i for $i \in [r]$. Note that M is also a matching in R^2. Let R be the restriction of R^2 to the vertices covered by M; then (R, M) is a matched $\mathcal{P}_R F$-partite k-system on U. For each edge $e \in M$, we arbitrarily relabel the clusters W_i for $i \in e$ as $V_1^e, \ldots, V_k^e$, and let $V^e = \bigcup_{j=1}^k V_j^e$. Then the following claim provides the partition of $V(J)$ required for the first step of the proof.

CLAIM 7.13. *There is a partition of $V(J)$ into an exceptional set Γ and sets X, Y, where X, Y are partitioned into X^e, Y^e for $e \in M$, and X^e, Y^e are partitioned into sets X_j^e, Y_j^e, $j \in [k]$, such that*

(i) *Γ is partitioned into sets $\Gamma_1, \ldots, \Gamma_t$ with $t \leq 3\psi n$, where each Γ_i has $b = k|I(F)|/r$ vertices in each V_i, $i \in [r]$,*
(ii) *for any sets $X_j^e \subseteq \Lambda_j^e \subseteq X_j^e \cup Y_j^e$, $j \in [k]$ with $|\Lambda_1^e| = \cdots = |\Lambda_k^e|$, writing $\Lambda^e = \Lambda_1^e \cup \cdots \cup \Lambda_k^e$, there is a perfect matching in $J'[\Lambda^e]$,*
(iii) *for any $j \in [r]$, $v \in V_j$, $\mathbf{i} \in I(F)$ with $i_j > 0$ and $S \subseteq V(J) \setminus X$ with $|S \cap V_i| \geq n/2 - n/6k$ for each i, there is an edge $e \in J[S \cup \{v\}]$ with $v \in e$ and $\mathbf{i}_\mathcal{P}(e) = \mathbf{i}$, and*
(iv) *$|X_j^e| = n_1/2$ and $|Y_j^e| = (1/2 - \psi/r)n_1$ for each $e \in M$ and $j \in [k]$.*

To prove the claim, we start by applying Proposition 6.9 and Theorem 6.4 to each $e \in M$, deleting at most θn_1 vertices from each V_j^e to obtain $\hat{V}_1^e, \ldots, \hat{V}_k^e$ and a k-partite k-complex G^e on $\hat{V}^e := \hat{V}_1^e \cup \cdots \cup \hat{V}_k^e$ such that

  (i) $G_k^e \subseteq G \setminus Z \subseteq J_k$,
  (ii) G^e is c-robustly 2^k-universal, and
  (iii) $|G_k^e(v)| \geq d^* n^{k-1}$ for any $v \in \hat{V}^e$.

For each $j \in [k]$ choose $X_j^e \subseteq \hat{V}_j^e$ of size $n_1/2$ uniformly at random and independently of all other choices. Then Claim 7.13 (ii) holds with probability $1 - o(1)$: this follows from Lemma 6.5, for which condition (i) is immediate, and condition (ii) holds by the following lemma from [**25**] (it is part of Lemma 4.4), which is proved by a martingale argument.

LEMMA 7.14. *Suppose that $1/n \ll d^*, b_2, 1/k, 1/b_1 < 1$. Let H be a k-partite k-graph with vertex classes $V_1, \ldots, V_k$, where $n \leq |V_i| \leq b_1 n$ for each $i \in [k]$. Also suppose that H has density $d(H) \geq d^*$ and that $b_2|V_i| \leq t_i \leq |V_i|$ for each i. If we choose a subset $X_i \subseteq V_i$ with $|X_i| = t_i$ uniformly at random and independently for each i, and let $X = X_1 \cup \cdots \cup X_k$, then the probability that $H[X]$ has density $d(H[X]) > d^*/2$ is at least $1 - 1/n^2$.*

Next, for each $e \in M$, arbitrarily delete up to $\psi n_1/r$ vertices from each $\hat{V}_j^e \setminus X_j^e$ to obtain Y_j^e such that $|X_j^e \cup Y_j^e| = (1 - \psi/r)n_1$ for each $j \in [k]$. Note that Claim 7.13 (iv) is then satisfied. Let Γ consist of all vertices of J which do not lie in some X^e or Y^e. Then (Γ, X, Y) is a partition of $V(J)$. Note that Γ consists of all vertices in clusters W_j which were not covered by edges of M, and all vertices deleted in forming the sets X^e and Y^e. There are at most $2\psi m_1 \cdot n_1 = 2\psi n$ vertices of the first type, and at most $rm_1 \cdot \psi n_1/r = \psi n$ vertices of the second type, so $|\Gamma| \leq 3\psi n$. Also, b divides $|\Gamma \cap V_i|$ for $i \in [r]$, as b divides n, and M is F-balanced. (We also use the fact that each $X_j^e \cup Y_j^e$, $e \in M$, $j \in [k]$ has the same size.) We fix an arbitrary partition of Γ into $\Gamma_1, \ldots, \Gamma_t$ that satisfies Claim 7.13 (i). It remains to satisfy Claim 7.13 (iii). Since $|X \cap V_i| \leq n/2$ for each i, applying the Chernoff bound, with probability $1 - o(1)$ we have $\delta_\ell^F(J[V(J) \setminus X]) \geq (k-\ell)n/3k$ for each $\ell \in [k-1]$ (we omit the calculation, which is similar to others in the paper, e.g. that in Theorem 7.10.) Now consider any j, v, $\mathbf{i}$ and S as in Claim 7.13 (iii). Starting with v, we greedily construct an edge $e \in J[S \cup \{v\}]$ with $\mathbf{i}_{\mathcal{P}}(e) = \mathbf{i}$ and $v \in e$. This is possible since $|X \cap V_i|, |S \cap V_i| > n/2 - n/6k$ for each i, so $\delta_\ell^F(J[V(J) \setminus X]) \geq n/3k > |V_i \setminus (X \cup S)|$ for $\ell \in [k-1]$ and $i \in [r]$. This completes the proof of Claim 7.13.

The second step of the proof is the following claim, which states that (R, M) contains small transferrals between any two vertices in any part of $\mathcal{P}_R$, even after deleting the vertices of a small number of edges of M.

CLAIM 7.15. *Let R' and M' be formed from R and M respectively by the deletion of the vertices covered by an F-balanced submatching $M_0 \subseteq M$ which satisfies $|M_0| \leq \alpha m$. Let $\mathcal{P}_{R'}$ be the partition of $V(R')$ into $U_1', \ldots, U_r'$ obtained by restricting $\mathcal{P}_R$. Then $D_C(R', M')[U_i']$ is complete for each $i \in [r]$.*

To prove the claim, we consider a matched k-system (R', M') formed in this manner and verify the conditions of Lemma 5.7. Note that each U_i' has size m', where $m \geq m' \geq (1 - k\alpha/r)m \geq (1 - 2k\alpha)m_1$, and that M' is F-balanced. Note also that if $e \in M'$, then $e \in M$, so $e \in R^1$. By Lemma 6.7 we therefore have $e \setminus \{v\} \in R^2$ for any $v \in e$; this was the reason for using two reduced systems. Since R' is the restriction of R^2 to the vertices of M', we therefore have $e \setminus \{v\} \in R'$ for any $v \in e$. Thus by Lemma 5.7, it suffices to verify the following conditions:

(a) $\delta^F(R') \geq \left(m', \left(\frac{k-1}{k} - \alpha'\right)m', \left(\frac{k-2}{k} - \alpha'\right)m', \ldots, \left(\frac{1}{k} - \alpha'\right)m'\right)$,

(b) for any $p \in [k-1]$ and sets $S_i \subseteq U_i'$ such that $|S_i| = pm'/k$ for each $i \in [r]$ there are at least $\beta(m')^{p+1}/10$ edges in $R_{p+1}'[S]$, where $S := \bigcup_{i \in [r]} S_i$, and

(c) $L_{\mathcal{P}_{R'}^*}(R_k')$ is complete with respect to $\mathcal{P}_{R'}$ for any partition $\mathcal{P}_{R'}^*$ of $V(R')$ which refines $\mathcal{P}_{R'}$ and whose parts each have size at least $m'/k - 2\alpha'm'$.

Condition (a) is immediate from (6), as R' was formed from R^2 by deleting at most $2k\alpha m_1$ vertices from each part including every vertex i for which $\{i\} \notin R^2$. For the same reason, since every edge of R^1 is also an edge of R^2, by Claim 7.12 we have (b). So it remains to verify condition (c). Consider any partition $\mathcal{P}_{R'}^*$ of $V(R')$ refining $\mathcal{P}_{R'}$ into d parts $U_1^*, \ldots, U_d^*$ each of size at least $m'/k - 2\alpha'm'$. Form a partition $\mathcal{P}_{R^2}^{**}$ of $[rm_1]$ into d parts $U_1^{**}, \ldots, U_d^{**}$, where for each $i \in [r]$ the at most $2k\alpha m_1$ vertices of $[rm_1] \setminus V(R')$ in the same part of $\mathcal{P}_{R^2}$ as U_i^* are inserted arbitrarily among the parts of $\mathcal{P}_{R'}^{**}$ in that part of $\mathcal{P}_{R^2}$. Then let $\mathcal{P}^\#$ be the partition of $V(J') = V(J)$ into d parts $U_j^\# := \bigcup_{i \in U_j^{**}} W_i$ for each j. Since any part of $\mathcal{P}_{R'}^*$ or $\mathcal{P}_{R^2}^{**}$ has size at least $m'/k - 2\alpha'm' \geq m_1/k - \mu m_1$, any part of $\mathcal{P}^\#$ has size at least $n/k - \mu n$. Then $L_{\mathcal{P}^\#}^\mu(J_k)$ is complete with respect to $\mathcal{P}$ by assumption (iii) on J. Now consider any $\mathbf{i}$ for which at least μn^k edges $e \in J_k$ have $\mathbf{i}_{\mathcal{P}^\#}(e) = \mathbf{i}$

(i.e. **i** is in the generating set of $L^\mu_{\mathcal{P}\#}(J_k)$). At least $\mu n^k/2$ edges $e \in J'_k$ then have $\mathbf{i}_{\mathcal{P}\#}(e) = \mathbf{i}$, and so by Lemma 6.10 it follows that at least $\mu m_1^k/4$ edges $e \in R_k^2$ have $\mathbf{i}_{\mathcal{P}^{**}_{R^2}}(e) = \mathbf{i}$. Since R' was formed from R^2 by deleting at most $2k\alpha m_1$ vertices from each part, there is at least one edge of R'_k with $\mathbf{i}_{\mathcal{P}^*_{R'}}(e) = \mathbf{i}$, so **i** is in the generating set of $L_{\mathcal{P}^*_{R'}}(R'_k)$. This verifies condition (c), so we have proved Claim 7.15.

Continuing with the proof of Theorem 7.11, the third step is to find a matching covering the exceptional set Γ. We proceed through $\Gamma_1, \ldots, \Gamma_t$ in turn, at each step covering a set Γ_i and using transferrals to rebalance the cluster sizes. Suppose we have chosen matchings $E_1, \ldots, E_{s-1}$ for some $s \in [t]$, where E_i covers the vertex set $V(E_i)$, with the following properties.

(a) The sets $V(E_i)$ for $i \in [s-1]$ are pairwise-disjoint and have size at most brk^2C,
(b) $\Gamma_i \subseteq V(E_i) \subseteq \Gamma_i \cup Y$ for each $i \in [s-1]$, and
(c) $|\bigcup_{i \in [s-1]} V(E_i) \cap Y_1^e| = \cdots = |\bigcup_{i \in [s-1]} V(E_i) \cap Y_k^e| \leq n_1/6k$ for any $e \in M$.

Then the following claim will enable us to continue the process.

CLAIM 7.16. *There is a matching E_s in J satisfying properties (a), (b), and (c) with s in place of $s-1$.*

To prove the claim, we need to first remove any clusters that have been too heavily used by $E_1, \ldots, E_{s-1}$. Note that

$$\sum_{i \in [s-1]} |V(E_i)| \leq brk^2Ct \leq 3brk^2C\psi n.$$

Thus there are at most $\alpha m_1/|I(F)|$ edges $e \in M$ such that $|\bigcup_{i \in [s-1]} V(E_i) \cap Y_j^e| \geq n_1/7k$ for some $j \in [k]$. Choose arbitrarily an F-balanced matching M_0 in R of size at most αm_1 which includes every such $e \in M$, and let (R', M') be the matched k-system obtained by deleting the vertices covered by M_0 from both R and M. Also let $\mathcal{P}_{R'}$ be the partition of $V(R')$ into $U'_1, \ldots, U'_r$ obtained by restricting $\mathcal{P}_R$. Then by Claim 7.15, $D_C(R', M')[U'_i]$ is complete for each $i \in [r]$. Let $Y' \subseteq Y$ consist of all vertices of Y except for those which lie in clusters deleted in forming R' and those which lie in some $V(E_i)$. Then for any $\ell \in [k]$ we have

$$|Y' \cap V_\ell| \geq |Y \cap V_\ell| - k\alpha m_1 n_1 - tbrk^2C \geq n/2 - n/8k.$$

Now choose for every $x \in \Gamma_s$ an edge $e_x \in J$ with $x \in e_x$ so that $\{e_x : x \in \Gamma_s\}$ is an F-balanced matching in J. Since $|\Gamma_s \cap V_i| = b = k|I(F)|/r$ for each i, by uniformity of F this will be accomplished by including k edges e_x of index **i** for each $\mathbf{i} \in I(F)$. By repeatedly applying Claim 7.13 (iii) we may choose these edges to lie within $Y' \cup \Gamma_s$ and to be pairwise-disjoint. The matching $\{e_x : x \in \Gamma_s\}$ then covers every vertex of Γ_s, but we may have unbalanced some clusters. To address this, consider the extra vertices $Q := \bigcup_{x \in \Gamma_s} e_x \setminus \{x\}$ that were removed. Since each $\mathbf{i} \in I(F)$ was represented k times, $Q_j := Q \cap V_j$ has size $(k-1)b$ for each $j \in [r]$. Divide each Q_j arbitrarily into b parts of size $k-1$, and for each $\mathbf{i} \in I(F)$ form $Q_\mathbf{i}$ of size $k(k-1)$ by taking i_j parts in Q_j, so that $(Q_\mathbf{i} : \mathbf{i} \in I(F))$ partitions Q and for each $\mathbf{i} \in I(F)$ the set $Q_\mathbf{i}$ has index $\mathbf{i}(Q_\mathbf{i}) = (k-1)\mathbf{i}$.

Next, for each $\mathbf{i} \in I(F)$ we arbitrarily pick a 'target' edge $e(\mathbf{i}) \in M'$ of index **i** to which we will transfer the imbalances caused by $Q_\mathbf{i}$. Then for any $\mathbf{i} \in I(F)$ and $x \in Q_\mathbf{i}$ we may choose $i(x) \in [m_1]$ and $i'(x) \in e(\mathbf{i})$ such that x is in the cluster $W_{i(x)}$, and $i(x)$ and $i'(x)$ are in the same part of $\mathcal{P}_{R'}$. Furthermore, we may choose the $i'(x)$'s so that the multiset $\{i'(x) : x \in Q_{\mathbf{i}}\}$ contains $k-1$ copies of each vertex of $e(\mathbf{i})$. Since $D_C(R', M')[U'_i]$ is complete for each $i \in [r]$, there is a simple $(i'(x), i(x))$-transferral of size at most C in (R', M'). Choose such a transferral for every $x \in Q$, and let (T, T') be the combination of these transferrals. To implement the transferral we need to select a matching $E_s^* = \{e^* : e \in T\}$ in $J[Y]$, whose edges correspond to the edges of T (counted with multiplicity), in that e^* contains one vertex in each cluster W_i with $i \in e$ for each $e \in T$. We can construct E_s^* greedily, using Lemma 6.1, since by Claim 7.13 (iv) Y has $n_1/2 - \psi n_1/r$ vertices in each cluster, and we have used at most $n_1/5k$ of these in $E_1, \ldots, E_{s-1}$ and edges so far chosen for E_s^*.

Now we let $E_s = \{e_x\}_{x \in \Gamma_s} \cup E_s^*$. Then $|E_s| = br + |T| \leq br + |Q|C \leq brkC$, so E_s covers at most brk^2C vertices. By construction E_s is disjoint from $E_1, \ldots, E_{s-1}$ and satisfies $\Gamma_s \subseteq V(E_s) \subseteq \Gamma_s \cup Y$. Furthermore, E_s avoids all clusters in which $\bigcup_{i \in [s-1]} V(E_i)$ covers at least $n_1/7k$ vertices. Thus it remains to show that $|V(E_s) \cap Y_1^e| = \cdots = |V(E_s) \cap Y_k^e|$ for any $e \in M$. To see this, note that the transferral (T, T') was chosen so that

$$\chi(T) - \chi(T') = \sum_{x \in Q} \chi(\{i'(x)\}) - \chi(\{i(x)\}) \in \mathbb{R}^{m_1}.$$

For each edge $e \in E_s$ we write $\chi^R(e) \in \mathbb{R}^{m_1}$ for the vector with $\chi^R(e)_i = |e \cap W_i|$. (Note that vertices in the exceptional set do not contribute here.) We also have $\sum_{e \in E_s^*} \chi^R(e) = \chi(T)$ and $\sum_{x \in \Gamma_s} \chi^R(e_x) = \sum_{x \in Q} \chi(\{i(x)\})$. It follows that

$$\sum_{e \in E_s} \chi^R(e) = \chi(T') + \sum_{x \in Q} \chi(\{i'(x)\}) = \chi(T') + \sum_{\mathbf{i} \in I(F)} (k-1)\chi(e(\mathbf{i})).$$

Since T' and $T' + (k-1)\sum_{\mathbf{i} \in I(F)} e(\mathbf{i})$ are multisets in M, this establishes the required property of E_s to prove Claim 7.16.

Thus we can greedily complete the third step of covering the exceptional set Γ, in such a way that for each $e \in M$ we use an equal number of vertices in each Y_j^e, $j \in [k]$. To finish the proof of the theorem, for each $e \in M$ and $j \in [k]$ let Λ_j^e consist of those vertices of $X_j^e \cup Y_j^e$ which are not covered by any of the sets $V(E_i)$ for $i \in [t]$. Then $X_j^e \subseteq \Lambda_j^e \subseteq X_j^e \cup Y_j^e$ and $|\Lambda_1^e| = \cdots = |\Lambda_k^e|$. By Claim 7.13 (ii), writing $\Lambda^e = \Lambda_1^e \cup \cdots \cup \Lambda_k^e$, there is a perfect matching in $J'_k[\Lambda^e]$. Combining these matchings for $e \in M$ and the matchings $E_1, \ldots, E_t$ we obtain a perfect matching M^* in J_k. Furthermore, all but at most $tbrkC \leq \gamma rn/kD_F^2$ edges of this matching lie in $J'[\Lambda^e]$ for some $e \in M$. Since M is F-balanced, it follows that M^* γ-represents F. □

We shall see shortly that Theorems 2.9 and 2.10 follow by a straightforward deduction from Theorem 7.11. However, the divisibility barriers considered there had the additional property of being transferral-free; we obtain this property using the next proposition.

PROPOSITION 7.17 *Let $1/n \ll \mu_1 \ll \mu \ll 1/r, 1/k$. Let $\mathcal{P}$ be a partition of a set V into parts $V_1, \ldots, V_r$ of size n, and J be a k-complex on V such that there exists a partition $\mathcal{P}_1$ of $V(J)$ which refines $\mathcal{P}$ into parts of size at least $n/k - \mu_1 n$ for which $L_{\mathcal{P}_1}^{\mu_1}(J_k)$ is incomplete with respect to $\mathcal{P}$. Then there there exists a partition*

$\mathcal{P}'$ of $V(J)$ which refines $\mathcal{P}$ into parts of size at least $n/k - \mu n$ such that $L^\mu_{\mathcal{P}'}(J_k)$ is transferral-free and incomplete with respect to $\mathcal{P}$.

PROOF. Introduce constants $\mu_1 \ll \cdots \ll \mu_{kr} = \mu$, and repeat the following step for $t \geq 1$. If $L^{\mu_t}_{\mathcal{P}_t}(J_k)$ is transferral-free, then terminate; otherwise $L^{\mu_t}_{\mathcal{P}_t}(J_k)$ contains some difference of index vectors $\mathbf{u}_i - \mathbf{u}_j$ with $i \neq j$. In this case, form a new partition $\mathcal{P}_{t+1}$ of V from $\mathcal{P}_t$ by merging parts V_i and V_j. Since $\mathcal{P}_1$ has at most kr parts, this process must terminate with a partition $\mathcal{P}' = \mathcal{P}_T$ for some $T \leq kr$. Observe that for any $t \geq 1$, if $L^{\mu_{t+1}}_{\mathcal{P}_{t+1}}(J_k)$ is complete with respect to $\mathcal{P}$ then the same must be true of $L^{\mu_t}_{\mathcal{P}_t}(J_k)$, since $L^{\mu_1}_{\mathcal{P}_1}(J_k)$ is incomplete with respect to $\mathcal{P}$ it follows that $L^{\mu_T}_{\mathcal{P}'}(J_k)$ is incomplete with respect to $\mathcal{P}$. Furthermore, the fact that the process terminated with $\mathcal{P}'$ implies that $L^{\mu_T}_{\mathcal{P}'}(J_k)$ is transferral-free; since $L^{\mu'}_{\mathcal{P}'}(J_k) \subseteq L^{\mu_T}_{\mathcal{P}'}(J_k)$ this completes the proof. □

Now, to deduce Theorem 2.9, we assume that properties 2 (Space barrier) and 3 (Divisibility barrier) do not hold, and show that property 1 (Matching) must hold. For this, introduce a new constant μ' with $\alpha\mu' \ll \mu$. We will apply Theorem 7.11 with $r = 1$ and F consisting of $(k!$ copies of) the unique function $f : [k] \to [1]$, so that $\delta^F(J) = \delta^f(J) = \delta(J)$ and $b = k|I(F)|/r = k$ divides n. Now the conditions of Theorem 7.11 are satisfied. Indeed, (i) holds by the minimum degree sequence, (ii) holds because there is no space barrier, and (iii) holds (with μ' in place of μ) by our assumption that there is no divisibility barrier combined with Proposition 7.17. Then J_k contains a perfect matching, so this proves Theorem 2.9.

Similarly, we can deduce Theorem 2.10 for the case where $b = k\binom{r}{k}/r = \binom{r-1}{k-1}$ divides n by taking F to consist of all injective functions $f : [k] \to [r]$; then $\delta^F(J) = \delta^*(J)$ is the partite minimum degree sequence, and the fact that M γ-represents F implies that M is γ-balanced. For the general case, write $d = \gcd(r, k)$, and note that k/d divides b and n, since we assume that $k \mid rn$. Thus we can choose $0 \leq a \leq b$ such that b divides $n - ak/d$. By choosing a matching with one edge of each index in $I = \{\sum_{i \in [k]} \mathbf{u}_{i+jd} : j \in [r/d]\}$ (where addition in the subscript is modulo r) we can remove k/d vertices from each part. We can delete the edges of a vertex-disjoint such matchings from J, which only slightly weakens the conditions of the theorem, so the general case follows from the case where $b \mid n$. We also note that we did not need the full strength of the degree assumption in Theorem 2.10 except to obtain that M is γ-balanced. For example, we could instead have taken the F generated by I as above, if $r > k$ is not divisible by k (so that F is connected).

Theorems 2.4 and 2.7 on matchings covering all but a constant number of vertices follow in the same way from the following common generalisation.

THEOREM 7.18. *Let $1/n \ll 1/\ell \ll \gamma \ll \alpha \ll \beta, \mu \ll 1/D_F, 1/r, 1/k$. Suppose F is a (k,r)-uniform connected allocation with $|F| \leq D_F$, and $\mathcal{P}$ partitions a set V into sets $V_1, \ldots, V_r$ each of size n. Suppose that J is a $\mathcal{P}F$-partite k-complex on V such that*

(i) $\delta^F(J) \geq \left(n, \left(\frac{k-1}{k} - \alpha\right)n, \left(\frac{k-2}{k} - \alpha\right)n, \ldots, \left(\frac{1}{k} - \alpha\right)n\right)$,
(ii) *for any $p \in [k-1]$ and sets $S_i \subseteq V_i$ such that $|S_i| = \lfloor pn/k \rfloor$ for each $i \in [r]$ there are at least βn^k edges of J'_k with more than p vertices in $S := \bigcup_{i \in [r]} S_i$, and*

Then J_k contains a matching which γ-represents F and covers all but at most ℓ vertices.

PROOF. The proof is very similar to that of Theorem 7.11, so we just indicate the necessary modifications. Take the same hierarchy of constants as in the proof of Theorem 7.11, and also take $\psi \ll \gamma \ll 1/B, 1/C \ll \alpha$. We start by forming the same partition of $V(J)$ as in Claim 7.13 (the divisibility conditions can be ensured by discarding a constant number of vertices). Instead of Claim 7.15, the corresponding claim here is that (R', M') is (B, C)-irreducible with respect to $\mathcal{P}_{R'}$. The proof is the same, except that we apply Lemma 5.6 instead of Lemma 5.7, and there is no lattice condition to check. Next, in the analogue of Claim 7.16, we can no longer ensure property (c), but instead we maintain the property

(c') $\quad \Big| \bigcup_{i \in [s]} V(E_i) \cap Y_j^e \Big| - \Big| \bigcup_{i \in [s]} V(E_i) \cap Y_{j'}^e \Big| \leq 4B$ for all $e \in M$ and $j, j' \in [k]$.

In constructing the sets E_s, we at first cover Γ_s by e_x, $x \in \Gamma_s$ but do not attempt to balance the cluster sizes. Instead, whenever we have formed a set E_s that would cause property (c') to fail, we remedy this using suitable b-fold transferrals with $b \leq B$. To do this, write

$$m_s^e = \sum_{j \in [k]} \Big| \bigcup_{i \in [s]} V(E_i) \cap Y_j^e \Big| \text{ and } d_s^{ej} = \Big| \bigcup_{i \in [s]} V(E_i) \cap Y_j^e \Big| - m_s^e / k$$

for $e \in M$, $j \in [k]$. Suppose we have formed E_s which has $d_s^{ej} > 2B$ for some $e \in M$ and $j \in [k]$. Let W be the cluster containing Y_j^e. Since each $\{e_x\}_{x \in \Gamma_s}$ uses the same number of vertices in each part V_i, $i \in [r]$, we can find another cluster W' in the same part as W, which contains some $Y_{j'}^{e'}$ for which $d_s^{e'j'} < 0$. By (B, C)-irreducibility there is a b-fold (W', W)-transferral (T, T') of size at most C, for some $b \leq B$. To implement the transferral, we include a matching in E_s^* whose edges correspond to the edges of T, using the same procedure as described in the proof of Theorem 7.11. Similarly, if $d_s^{ej} < -2B$ for some $e \in M$ and $j \in [k]$, then we can find another cluster W' in the same part as W which contains some $Y_{j'}^{e'}$ for which $d_s^{e'j'} > 0$. We then implement a b-fold (W, W')-transferral (T, T') of size at most C, for some $b \leq B$, in the same way as before. By repeating this process while there is any $d_s^{ej} > 2B$ for some $e \in M$ and $j \in [k]$, we can construct E_s^* which when added to E_s satisfies property (c'). Thus we can greedily cover the exceptional set Γ, in such a way that for each $e \in M$ and $j, j' \in [k]$ the number of vertices used from Y_j^e and $Y_{j'}^e$ differ by at most $4B$. Now we discard at most $4B$ vertices arbitrarily from each Y_j^e to balance the cluster sizes, so that Claim 7.13 (ii) gives a perfect matching on the remaining sets. Combining all the matchings, we have covered all vertices, except for some number that is bounded by a constant independent of n. □

CHAPTER 8

Packing Tetrahedra

In this chapter we prove Theorem 1.1, which determines precisely the codegree threshold for a perfect tetrahedron packing in a 3-graph G on n vertices, where $4 \mid n$ and n is sufficiently large. The theorem states that if $8 \mid n$ and $\delta(G) \geq 3n/4 - 2$, or if $8 \nmid n$ and $\delta(G) \geq 3n/4 - 1$, then G contains a perfect K_4^3-packing. We start with a construction due to Pikhurko [44] showing that this minimum degree bound is best possible; the complement of this construction is illustrated in Figure 4.

PROPOSITION 8.1. *For any n divisible by 4 there exists a 3-graph G on n vertices with*
$$\delta(G) \geq \begin{cases} 3n/4 - 3 & \text{if } 8 \mid n \\ 3n/4 - 2 & \text{otherwise,} \end{cases}$$
which does not contain a perfect K_4^3-packing.

PROOF. Partition a set V of n vertices into parts V_1, V_2, V_3, V_4 such that $|V_1|$ is odd and the set sizes $|V_j|$ are as equal as possible. So $|V_j| \leq n/4 + 1$ for every $j \in [4]$, and if $8 \nmid n$ then $|V_j| = n/4$ for each n. Let G be the 3-graph on vertex set V whose edges are all $e \in \binom{V}{3}$ except for those e such that
  (i) $|e \cap V_1| = 3$,
  (ii) $|e \cap V_1| = 1$ and $|e \cap V_i| = 2$ for some $i \in \{2, 3, 4\}$, or
  (iii) $|e \cap V_2| = |e \cap V_3| = |e \cap V_4| = 1$.
To calculate $\delta(G)$, consider any pair of vertices xy. We claim that there is precisely one class V_i, $i \in [4]$ such that xyz is not an edge for $z \in V_i$. For this we just check all cases: if x, y are in the same class then $i = 1$, if $x \in V_1$ and $y \in V_j$ with $j \neq 1$ then $i = j$, and if x, y are in different parts among $\{V_2, V_3, V_4\}$ then V_i is the third of these parts. Thus $\delta(G) = n - 2 - \max\{|V_1|, |V_2|, |V_3|, |V_4|\}$, which has value $3n/4 - 3$ if $8 \mid n$ and $3n/4 - 2$ otherwise. Furthermore, we claim that any copy X of K_4^3 in G must have an even number of vertices in V_1. To see this, note first that we cannot have $|X \cap V_1| = 3$, since there are no edges contained in V_1. Now suppose that $|X \cap V_1| = 1$. Of the remaining 3 vertices, either 2 lie in the same part among $\{V_2, V_3, V_4\}$, or all 3 parts are represented; either way some triple in X is not an edge of G. Therefore $|X \cap V_1|$ is even. Since $|V_1|$ is odd, we conclude that G does not contain a perfect K_4^3-packing. □

Now we start on the proof of Theorem 1.1, which proceeds as follows. We will apply our results on perfect matchings in simplicial complexes to find a perfect matching in the clique 4-complex $J = J_4(G)$. Note that this is precisely a perfect tetrahedron packing in G. We shall see in the next section that J satisfies the usual minimum degree assumption, and so we consider the possible space and divisibility barriers to such a perfect matching. One such barrier is the divisibility barrier in the 4-graph constructed in Proposition 8.1. Indeed, in this construction every edge

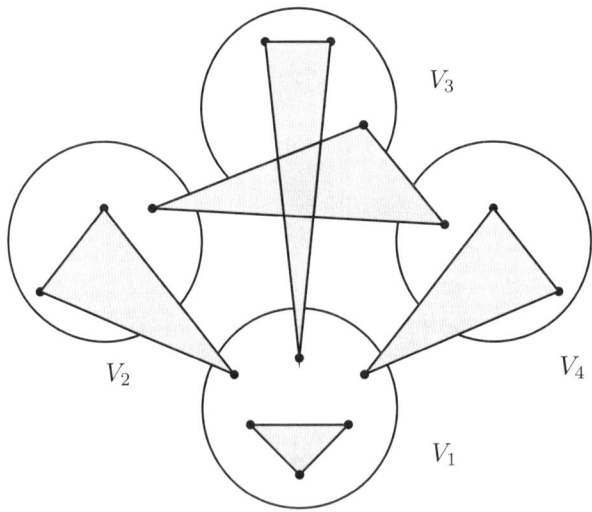

FIGURE 4. The 3-graph G formed in Proposition 8.1 consists of all edges except for those of the types shown above. Note that any copy of K_4^3 in G must have an even number of vertices in V_1.

of J_4 (that is, every tetrahedron) has even intersection with V_1. In fact, we shall see that this is the only space or divisibility barrier to a perfect matching in J_4 (that is, a tetrahedron packing in G).

In Section 8.1, we show that there is no space barrier to a perfect matching in G. Furthermore, this is true even under the weaker assumption $\delta(G) \geq (3/4 - c)n$ for some absolute constant $c > 0$. In particular, it then follows from Theorem 2.4 that G contains a tetrahedron packing that covers all but at most ℓ vertices, for some absolute constant ℓ, even under this weaker assumption. Thus we demonstrate that the threshold for covering all but a constant number of vertices is asymptotically different from that for covering all vertices.

Next, in Section 8.2 we prove some results on how the densities of edges in a k-complex J of different index vectors are related to one another. Indeed, for a fixed partition of $V(J)$ we define the density $d_{\mathbf{i}}(G)$ to be the proportion of possible edges of G of index $\mathbf{i}$ which are in fact actual edges of G. Then, for example, if $V(G)$ is partitioned into parts V_1, V_2 and V_3, we should be able to say that if $d_{(1,1,2)}$ is large, then $d_{(1,1,1)}$, $d_{(0,1,2)}$ and $d_{(1,0,2)}$ should each be large (since G is a k-complex). Similarly, if $d_{(1,1,1)}$ is large, and G has high codegree, then at least one of $d_{(2,1,1)}$, $d_{(1,2,1)}$ and $d_{(1,1,2)}$ should be large. In Section 8.2 we prove general forms of these and other arguments, which we make extensive use of in Sections 8.3 and 8.4.

We then proceed to the most difficult part of the proof, namely the analysis of divisibility barriers, for which we must consider Turán-type problems for K_4^3 (fortunately we can solve these ones, unlike Turán's famous conjecture!) If we do not have a perfect matching in J_4, then having shown that space barriers are impossible, we conclude from Theorem 2.9 that there must be a divisibility barrier. In Section 8.3 we consider divisibility barriers with two parts, and then in Section 8.4 we consider divisibility barriers with more than two parts. Taken together, in these sections we show that all but one of the potential divisibility barriers are impossible, and furthermore that the possible divisibility barrier implies a stability result, that G is structurally close to the construction described in Proposition 8.1. This sets up an application of the stability method in Section 8.5: either G is not structurally close to the construction, in which case Theorem 2.9 gives a perfect matching in J_4, or G is structurally close to the construction, in which case we give a separate argument exploiting this structure to see that again we have a perfect matching in J_4, that is, a perfect tetrahedron packing in G.

8.1. Packing to within a constant

As a prelude to our result on perfect tetrahedron packings, we prove the following result that demonstrates that there is a different threshold for packing to within a constant; the ingredients of its proof will also be used for the main result.

THEOREM 8.2. *Suppose $1/n \ll 1/\ell \ll c \ll 1$ and G is a 3-graph on n vertices with $\delta(G) \geq (3/4 - c)n$. Then G contains a tetrahedron packing that covers all but at most ℓ vertices.*

A reformulation of the desired conclusion is that there is a matching covering all but at most ℓ vertices in J_4, where $J = J_4(G)$ is the clique 4-complex. First we express the minimum degree sequence of J in terms of the minimum degree of G. Since J_i is complete for $i \leq 2$, we have $\delta_0(J) = n$, $\delta_1(J) = n-1$ and $\delta_2(J) = \delta(G)$. For $\delta_3(J)$, consider any $e = xyz \in J_3 = G$. For each of the pairs xy, xz, yz, there are at most $(n-3) - (\delta(G)-1)$ vertices $a \in V(G) \setminus \{x,y,z\}$ that do not form an edge of G with that pair. Thus $xyza$ is a tetrahedron in G for at least $(n-3) - 3(n-\delta(G)-2) = 3\delta(G) - 2n + 3$ vertices a. Therefore
(8) $$\delta(J_4(G)) \geq (n, n-1, \delta(G), 3\delta(G) - 2n + 3).$$

With our current assumption $\delta(G) \geq (3/4 - c)n$, this gives $\delta(J) \geq (n, n-1, (3/4-c)n, (1/4-3c)n + 3)$, so our usual minimum degree sequence assumption is satisfied; in fact, the bounds on $\delta_1(J)$ and $\delta_2(J)$ happen to be much stronger than necessary. Next, we need to consider space barriers, of which there are 3 possible types, corresponding to sets S of size $n/4$, $n/2$ or $3n/4$. The first two cases are covered by the next proposition, which is a straightforward application of the minimum degree sequence.

PROPOSITION 8.3. *Suppose $c \ll 1$ and G is a 3-graph on n vertices with $\delta(G) \geq (3/4 - c)n$. Then*

(i) for any $S \subseteq V(J)$ of size $n/4$, the number of edges of J_4 with at least 2 vertices in S is at least $n^4/3000$, i.e. J_4 is not $1/3000$-contained in $J(S,1)_4$, and

(ii) for any $S \subseteq V(J)$ of size $n/2$, the number of edges of J_4 with at least 3 vertices in S is at least $n^4/2000$, i.e. J_4 is not $1/2000$-contained in $J(S,2)_4$.

PROOF. For (i), we construct edges $v_1v_2v_3v_4$ with $v_1, v_2 \in S$. There are $n/4$ choices for v_1, $n/4 - 1$ choices for v_2, at least $(3/4 - c)n$ choices for v_3, and at least $(1/4 - 3c)n$ choices for v_4. Each such edge is counted at most 24 times, so the bound follows. Similarly, for (ii) we construct edges $v_1v_2v_3v_4$ with $v_1, v_2, v_3 \in S$. There are $n/2$ choices for v_1, $n/2 - 1$ choices for v_2, at least $|S| - (n - \delta(G)) \geq (1/4 - c)n$ choices for v_3, and at least $(1/4 - 3c)n$ choices for v_4, so again the bound follows. □

It remains to consider the case $|S| = 3n/4$, for which we need to show that many edges of J_4 (that is, many tetrahedra in G) are contained in S. This follows from known bounds on the Turán density of K_4^3 and 'supersaturation'. Here we only quote what suffices for our purposes, so we refer the reader to the survey [23] for more information. Given an r-graph F, the Turán number $\mathrm{ex}(n, F)$ is the maximum number of edges in an F-free r-graph on n vertices. The Turán density is $\pi(F) := \lim_{n \to \infty} \binom{n}{r}^{-1} \mathrm{ex}(n, F)$. We use the following result of Chung and Lu [5].

THEOREM 8.4. $\pi(K_4^3) \leq (3 + \sqrt{17})/12 (\approx 0.5936)$.

We combine this with the following supersaturation result of Erdős and Simonovits [13].

THEOREM 8.5. *For any r-graph F and $a > 0$ there are $b, n_0 > 0$ so that if G is an r-graph on $n > n_0$ vertices with $e(G) > (\pi(F) + a)\binom{n}{r}$ then G contains at least $bn^{|V(F)|}$ copies of F.*

Given a set $S \subseteq V(J)$ of size sn, any pair xy in S is contained in at least $\delta(G) - (1 - s)n$ edges of $G[S]$. Under the assumption $\delta(G) \geq (3/4 - c)n$, we get $|G[S]| \geq \frac{1}{3}\binom{|S|}{2}(s - 1/4 - c)n \geq (1 - 1/4s - c/s)\binom{|S|}{3}$. Note that if $s \geq 5/8$ then $1 - 1/4s \geq 3/5$, so $1 - 1/4s - c/s > \pi(K_4^3) + 1/200$ for small c. Then Theorems 8.4 and 8.5 imply the following result.

PROPOSITION 8.6. *Suppose $1/n \ll b, c \ll 1$ and G is a 3-graph on n vertices with $\delta(G) \geq (3/4 - c)n$. Then for any $S \subseteq V(J)$ with $|S| \geq 5n/8$, the number of edges of J_4 contained in S is at least bn^4. In particular, if $|S| = 3n/4$ then J_4 is not b-contained in $J(S, 3)_4$.*

Now Theorem 8.2 follows by applying Theorem 2.4 to the clique 4-complex $J_4(G)$. Indeed, the degree sequence assumption holds by (8), and there are no space barriers by Propositions 8.3 and 8.6. Thus there is a matching covering all but at most ℓ vertices in $J_4(G)_4$, as required.

8.2. Properties of index vectors

Now we need some notation and simple properties of index vectors. Let X be a set which is partitioned into r non-empty parts $X_1, \ldots, X_r$. We let $K(X)$ denote the complete k-complex on X, where any $e \in \binom{X}{\leq k}$ is an edge of $K(X)$. Recall that the index vector $\mathbf{i}(e) \in \mathbb{Z}^r$ of a set $e \subseteq X$ has coordinates $i(e)_j = |e \cap X_j|$. If $\mathbf{i}(e) = \mathbf{i}$, then the possible index vectors of subsets of e are precisely those $\mathbf{i}' \in \mathbb{Z}^r$ with $\mathbf{0} \leq \mathbf{i}' \leq \mathbf{i}$, where as usual $\leq$ is the pointwise partial order on vectors. Also recall that if J is a k-system on X, then $J_\mathbf{i}$ denotes the set of edges in J with index $\mathbf{i}$. We define the *density of J at $\mathbf{i}$* as

$$d_\mathbf{i}(J) := \frac{|J_\mathbf{i}|}{|K(X)_\mathbf{i}|}.$$

Note that this notation should not be confused with that for the relative density of a complex used in Chapter 6 (which we will not use in this chapter). Similarly to the notation used for complexes, we let $J_\mathbf{i}^*$ denote the set of $e \in K(X)_\mathbf{i}$ such that $e' \in J$ for every strict subset $e' \subset e$. We also recall that we write $|\mathbf{i}| = \sum_{j \in [r]} i_j$, and that $\mathbf{u}_i$ denotes the standard basis vector with ith coordinate 1 and all other coordinates 0. Given an index vector $\mathbf{i} = (i_1, \ldots, i_r)$, we write $\partial \mathbf{i}$ for the multiset that contains i_j copies of $\mathbf{i} - \mathbf{u}_j$ for each $j \in [r]$. The next proposition sets out some useful properties linking the densities of different indices.

PROPOSITION 8.7. *Let J be a k-complex on a set X partitioned as $(X_1, \ldots, X_r)$. Then*

(i) *for any $\mathbf{i}, \mathbf{i}'$ with $\mathbf{i}' \leq \mathbf{i}$ we have $d_{\mathbf{i}'}(J) \geq d_{\mathbf{i}}(J)$,*
(ii) *for any $\mathbf{i}$ we have $\sum_{j \in [r]} d_{\mathbf{i}+\mathbf{u}_j}(J)|X_j| \geq \delta_{|\mathbf{i}|}(J) d_{\mathbf{i}}(J)$,*
(iii) *if $J_{\mathbf{i}} = J_{\mathbf{i}}^*$ for some $\mathbf{i} = (i_1, \ldots, i_r)$, then $d_{\mathbf{i}}(J) \geq 1 - \sum_{j \in [r]} i_j(1 - d_{\mathbf{i}-\mathbf{u}_j}(J))$,*
and
(iv) *for any $\mathbf{i}$ we have $\prod_{\mathbf{i}' \in \partial \mathbf{i}} d_{\mathbf{i}'}(J) \geq d_{\mathbf{i}}(J)^{|\mathbf{i}|-1} + O(1/|X|)$.*

PROOF. For (i), it is sufficient to consider the case when $|\mathbf{i}'| = |\mathbf{i}| - 1$. We can write $\mathbf{i}' = \mathbf{i} - \mathbf{u}_j$ for some $j \in [r]$ such that $i_j \geq 1$. The required density inequality is a variant of the Local LYM inequality (see e.g. [**3**, Theorem 3.3]). We briefly give the standard double-counting argument, which is as follows. Consider the pairs (e', e) with $e' \in J_{\mathbf{i}'}$, $e \in J_{\mathbf{i}}$ and $e' \subseteq e$. The number of such pairs is at least $i_j|J_{\mathbf{i}}|$, and at most $(|X_j| - i_j + 1)|J_{\mathbf{i}'}|$. It follows that

$$\frac{|J_{\mathbf{i}'}|}{|J_{\mathbf{i}}|} \geq \frac{i_j}{(|X_j| - i_j + 1)} = \binom{|X_j|}{i_j - 1} \Big/ \binom{|X_j|}{i_j} = \frac{|K(X)_{\mathbf{i}'}|}{|K(X)_{\mathbf{i}}|},$$

which gives (i).

For (ii), we consider the pairs (e, v) with $e \in J_{\mathbf{i}}$ and $e \cup \{v\} \in J$. The number of such pairs is at least $\delta_{|\mathbf{i}|}(J)|J_{\mathbf{i}}|$, and at most $\sum_{j \in [r]}(i_j + 1)|J_{\mathbf{i}+\mathbf{u}_j}|$. This gives the stated inequality, using $|J_{\mathbf{i}}| = d_{\mathbf{i}}(J)|K(X)_{\mathbf{i}}|$, $|J_{\mathbf{i}+\mathbf{u}_j}| = d_{\mathbf{i}+\mathbf{u}_j}(J)|K(X)_{\mathbf{i}+\mathbf{u}_j}|$, and $|X_j||K(X)_{\mathbf{i}}| = (i_j + 1)|K(X)_{\mathbf{i}+\mathbf{u}_j}|$.

For (iii), choose e in $K(X)_{\mathbf{i}}$ uniformly at random. For each $e' \subset e$ with $|e'| = |e|-1$, writing $\mathbf{i}(e') = \mathbf{i}-\mathbf{u}_j$, the probability that e' is *not* an edge of J is $1-d_{\mathbf{i}-\mathbf{u}_j}(J)$. Thus the probability that every strict subset $e' \subset e$ is an edge of J is at least $1 - \sum_{j \in [r]} i_j(1 - d_{\mathbf{i}-\mathbf{u}_j}(J))$. Since $J_{\mathbf{i}} = J_{\mathbf{i}}^*$, this event implies that $e \in J$, so we have the stated density.

For (iv), we first consider the multipartite case, when each i_j is 0 or 1, and let $I = \{j : i_j = 1\}$. We apply Shearer's Lemma (see [**4**]), which states that if $\mathcal{A}$ and $\mathcal{F}$ are families of subsets of S such that each element of S is contained in at least ℓ elements of $\mathcal{A}$, and $\mathcal{F}_A = \{A \cap F : F \in \mathcal{F}\}$ for $A \in \mathcal{A}$, then $\prod_{A \in \mathcal{A}} |\mathcal{F}_A| \geq |\mathcal{F}|^\ell$. Here we take $S = X_I$, $\mathcal{F} = J_{\mathbf{i}}$, and $\mathcal{A} = \{X_{I \setminus j}\}_{j \in I}$, so $\ell = |\mathbf{i}| - 1$. Then $\prod_{j \in I} |J_{\mathbf{i}-\mathbf{u}_j}| \geq |J_{\mathbf{i}}|^{|\mathbf{i}|-1}$, which gives the result (even without the $O(1/|X|)$ term). For the general case we apply the multipartite case to an auxiliary k-partite k-complex H, whose parts $X_1', \ldots, X_{|\mathbf{i}|}'$ consist of i_j copies of X_j for $j \in [r]$, and whose edges are all k-partite copies of each edge in $\bigcup_{\mathbf{i}' \leq \mathbf{i}} J_{\mathbf{i}'}$ (so, for example, each edge of $J_{\mathbf{i}}$ corresponds to $\prod_{j \in [r]} i_j!$ of H_k). Then $d_{\mathbf{1}}(H) = d_{\mathbf{i}}(J) + O(1/|X|)$ and $\{d_{\mathbf{1}-\mathbf{u}_j}(H)\}_{j \in [|\mathbf{i}|]}$ has i_j copies of $d_{\mathbf{i}-\mathbf{u}_j}(J) + O(1/|X|)$ for $j \in [r]$, so the result follows. □

8.3. Divisibility barriers with two parts

For the perfect packing result, we need to consider divisibility barriers. We start with the case of a bipartition, for which there are two natural candidates for a barrier: (i) almost all tetrahedra are 'even', i.e. have an even number of vertices in both parts, and (ii) almost all tetrahedra are 'odd', i.e. have an odd number of vertices in both parts. We start by showing that these are indeed the only two possibilities. Note that here, and for most of the proof, we only assume the weaker degree hypothesis used in the previous section; the exact degree condition is only used in Lemma 8.10.

LEMMA 8.8. *Suppose $1/n \ll \mu, c \ll 1$ and G is a 3-graph on n vertices with $\delta(G) \geq (3/4 - c)n$. Write $J = J_4(G)$ for the clique 4-complex of G. Suppose $\mathcal{P}$ is a partition of V into two parts V_1, V_2 of size at least $n/4 - \mu n$ such that $L_{\mathcal{P}}^{\mu}(J_4)$ is incomplete. Then either (i) there are fewer than $3\mu n^4$ odd tetrahedra, or (ii) there are fewer than $3\mu n^4$ even tetrahedra.*

PROOF. Suppose for a contradiction that there are at least $3\mu n^4$ odd tetrahedra and at least $3\mu n^4$ odd tetrahedra. Then $L := L_{\mathcal{P}}^{\mu}(J_4)$ contains at least one of $(1,3)$ and $(3,1)$, and also at least one of $(4,0)$, $(2,2)$ and $(0,4)$. By (8), we have $\delta(J) \geq (n, n-1, (3/4-c)n, (1/4-3c)n)$. Next we apply Proposition 8.7(ii) to $(1,1)$, which gives
$$d_{(2,1)}(J)|V_1| + d_{(1,2)}(J)|V_2| \geq (3/4 - c)n,$$
and to $(2,0)$, which gives
$$d_{(3,0)}(J)|V_1| + d_{(2,1)}(J)|V_2| \geq (3/4 - c)n.$$
Summing the inequalities we deduce
$$d_{(2,1)}(J)n \geq (3/2 - 2c)n - |V_2| - |V_1| = (1/2 - 2c)n.$$
Now applying Proposition 8.7(ii) to $(2,1)$ we get
$$d_{(3,1)}(J)|V_1| + d_{(2,2)}(J)|V_2| \geq (1/2 - 2c)(1/4 - 3c)n,$$
so L contains at least one of $(3,1)$ and $(2,2)$. Similarly, L contains at least one of $(1,3)$ and $(2,2)$. Now we can deduce that $(-1,1) \in L$. Indeed, if $(2,2) \in L$ this holds since $(-1,1) = (1,3) - (2,2) = (2,2) - (3,1)$ and L contains at least one of $(1,3)$ and $(3,1)$. On the other hand, if $(2,2) \notin L$ then L contains both $(3,1)$ and $(1,3)$. Also, L contains at least one of $(4,0)$ and $(0,4)$. Since $(-1,1) = (3,1) - (4,0) = (0,4) - (1,3)$, again we get $(-1,1) \in L$. However, this contradicts the assumption that L is incomplete. $\square$

Next we show that it is impossible to avoid even tetrahedra.

LEMMA 8.9. *Suppose $1/n \ll b, c \ll 1$ and G is a 3-graph on n vertices with $\delta(G) \geq (3/4 - c)n$. Then for any bipartition of $V(G)$, there are at least bn^4 copies of K_4^3 with an even number of vertices in both parts.*

PROOF. Consider any partition (A, B) of $V(G)$. Form a digraph D on vertex set $V(D) := \binom{A}{2} \cup \binom{B}{2}$, where there is an edge from $P \in V(D)$ to $Q \in V(D)$ precisely if $P \cup \{q\}$ is an edge of G for each $q \in Q$. Then there is a one-to-one correspondence between 2-cycles $\{PQ, QP\}$ in D and copies of K_4^3 in G with an even number of vertices in A. We write $|A| = an$, where without loss of generality we have $0 \leq a \leq 1/2$. We can also assume that $a \geq 3/8$, as otherwise $|B| \geq 5n/8$, so the desired conclusion follows from Proposition 8.6. Then by convexity,
$$|V(D)| = \binom{|A|}{2} + \binom{|B|}{2} \leq \binom{3n/8}{2} + \binom{5n/8}{2} \leq \frac{17n^2}{64}.$$
Also, for any $P \in V(D)$, we can write $d_D^+(P) = \binom{d_A(P)}{2} + \binom{d_A(B)}{2}$, where $d_X(P)$ is the number of vertices $x \in X$ such that $P \cup \{x\} \in G$, for $X \in \{A, B\}$. Since $d_A(P) + d_B(P) = \delta(G) \geq (3/4 - c)n$, by convexity
$$\delta^+(D) \geq 2\binom{(3/4-c)n/2}{2} > \frac{9n^2}{64} - cn^2.$$

So the number of 2-cycles in D is at least $(9|V(D)|/17-cn^2)|V(D)|-\binom{|V(D)|}{2} \geq bn^4$, as required. $\square$

Now we consider odd tetrahedra. Recall that the construction in Proposition 8.1 gave a 3-graph G on n vertices partitioned into two parts V_1 and $V_2 \cup V_3 \cup V_4$, so that $|V_1|$ is an odd number close to $n/4$, and all tetrahedra are even with respect to the bipartition. The next lemma shows that the minimum degree condition of Theorem 1.1 is sufficient to prevent any such partition in G. That is, for any k-graph G which meets the conditions of Theorem 1.1, and any partition of $V(G)$ into two not-too-small parts, there is some tetrahedron in H which intersects each part in an odd number of vertices. This is the only part of the proof of Theorem 1.1 for which the minimum degree bound is tight.

LEMMA 8.10. *Suppose G is a 3-graph on n vertices with $4 \mid n$ and $\delta(G) \geq 3n/4 - 1$. Let (A,B) be a partition of $V(G)$ with $3n/16 < |A| < 5n/16$. Then G contains a copy of K_4^3 with an odd number of vertices in both parts.*

PROOF. Consider such a partition (A, B), and suppose for a contradiction that there is no odd tetrahedron. By (8), for any edge $e \in G$, the number of vertices v such that $e \cup \{v\}$ induces a copy of K_4^3 is at least $3\delta(G) - 2n + 3 \geq n/4$. If e has an odd number of vertices in A, then each such vertex v must also lie in A. Consider any $u \in A$ and $v \in B$. Since $\delta(G) > n - |B|$ we may choose $v' \in B$ such that $e = uvv'$ is an edge of G. Then e has an odd number of vertices in A, so there must be at least $n/4$ vertices $u' \in A$ such that $e \cup \{u'\}$ induces a copy of K_4^3. Each such u' is distinct from u, so we must have $|A| \geq n/4 + 1$.

Now consider any distinct $x, y \in B$. There are at least $\delta(G) \geq 3n/4 - 1$ vertices $z \in V(G)$ such that $xyz \in G$. At most $|B| - 2 \leq 3n/4 - 3$ of these lie in B, so we may choose some such $z \in A$. Then $e' = xyz$ has an odd number of vertices in A, so again there must be at least $n/4$ vertices $w \in A$ such that $e' \cup \{w\}$ induces a copy of K_4^3 in G. But x and y were arbitrary, so we deduce that for any $x, y \in B$, there are at most $|A| - n/4 < n/16$ vertices $z \in A$ for which xyz is not an edge of G. Since $|A| > 3n/16$, for any edge $x'y'z'$ in $G[B]$, there is some $u' \in A$ such that $u'x'y'z'$ induces a copy of K_4^3 in G, which contradicts the assumption that there is no odd tetrahedron. $\square$

The previous lemma will be useful when considering the extremal case, but a single odd tetrahedron does not suffice to rule out a divisibility barrier. To find many odd tetrahedra we need a result relating the triangle density to the edge density in any graph. Given a graph G on n vertices, the edge density of G is $d(G) = e(G)\binom{n}{2}^{-1}$, and the triangle density is $\triangle(G) = N_3(G)\binom{n}{3}^{-1}$, where $N_3(G)$ is the number of triangles in G. In the following theorem, the first part follows from a result of Goodman [16], and the second from a result of Lovász and Simonovits [37, Theorem 2]. (Razborov [45] proved an asymptotically tight general result.)

THEOREM 8.11. *Suppose G is a graph on n vertices. Then*
(i) $\triangle(G) \geq d(G)(2d(G) - 1)$, and
(ii) if $1/n \ll \varepsilon \ll \varepsilon' \ll 1$, $d(G) = 2/3 \pm \varepsilon$ and $\triangle(G) = 2/9 \pm \varepsilon$, then we may partition $V(G)$ into sets V_1, V_2, V_3 each of size at least $n/3 - \varepsilon'n$ so that $d(G[V_i]) \leq \varepsilon'$ for each $i \in [3]$.

Now we can prove the following result, which states that, even under the weaker degree assumption $\delta(G) \geq (3/4 - c)n$, either there is no divisibility barrier for odd tetrahedra, or G is close to the construction of Proposition 8.1. This lemma plays a key role in the stability argument outlined at the start of this chapter. Indeed, to prove Theorem 1.1 we will consider first two cases: either there is no divisibility barrier, in which case Theorem 2.9 implies that $J_4(G)_4$ contains a perfect matching, or there is a divisibility barrier, in which case the structure provided by Lemma 8.12, combined with the odd tetrahedron guaranteed by Lemma 8.10, will enable us to construct a perfect matching in $J_4(G)_4$.

Let $\mathcal{P}$ partition a set V into parts V_1, V_2, V_3, V_4. We say that an index vector $\mathbf{i}$ with respect to $\mathcal{P}$ is *bad* if it is not used by edges of the extremal example, i.e. $\mathbf{i}$ is one of $(3,0,0,0)$, $(1,2,0,0)$, $(1,0,2,0)$, $(1,0,0,2)$ or $(0,1,1,1)$; otherwise $\mathbf{i}$ is *good*.

LEMMA 8.12. *Let $1/n \ll b, c \ll \gamma \ll 1$ and G be a 3-graph on n vertices with $\delta(G) \geq (3/4 - c)n$. Suppose that there is some partition of $V(G)$ into parts U_1 and U_2 each of size at least $n/4 - bn$ such that at most bn^4 copies of K_4^3 in G are 'odd', i.e. have an odd number of vertices in each part. Then there is a partition of $V(G)$ into parts V_1, V_2, V_3, V_4 each of size at least $n/4 - \gamma n$, such that V_1 is equal to one of U_1 or U_2, and $d_{\mathbf{i}}(G) \geq 1 - \gamma$ for each good $\mathbf{i}$.*

PROOF. Introduce new constants with $b, c \ll \alpha \ll \beta \ll \beta' \ll \gamma$. Let $J = J_4(G)$ be the clique 4-complex of G. By (8), we have $\delta(J) \geq (n, n-1, (3/4-c)n, (1/4-3c)n)$. Write

$$\tau := d_{(3,0)}(J) = d(G[U_1]), \qquad \tau' := d_{(0,3)}(J) = d(G[U_2])$$
$$\rho := d_{(2,1)}(J) - \alpha, \qquad \rho' := d_{(1,2)}(J) - \alpha.$$

Let Z be the set of vertices that lie in more than $\sqrt{b}n^3$ odd tetrahedra. Then $|Z| < 4\sqrt{b}n$. Let $U_1' = \{v \in U_1 : d(G(v)[U_2]) \geq \rho'\}$. Then

$$\rho' + \alpha = d_{(1,2)}(J) = \frac{1}{|U_1|} \sum_{v \in U_1} d(G(v)[U_2]) \leq \frac{|U_1'| + |U_1 \setminus U_1'|\rho'}{|U_1|},$$

so $|U_1'| \geq \alpha|U_1|$. Thus we can choose $v \in U_1' \setminus Z$. We claim that $G(v)[U_2]$ must have triangle density at most $1 - \tau' + \alpha$. Otherwise, we would have at least $\alpha\binom{|U_2|}{3} > \sqrt{b}n^3$ triples in U_2 that are both triangles in $G(v)[U_2]$ and edges of $G[U_2]$. But these triples form odd tetrahedra with v, which contradicts $v \notin Z$. So $G(v)[U_2]$ has triangle density at most $1 - \tau' + \alpha$, and, since $v \in U_1'$, edge density at least ρ'. Now Theorem 8.11(i) gives

(9) $$\rho'(2\rho' - 1) + \tau' - \alpha \leq 1.$$

The same argument with the roles of U_1 and U_2 switched shows that

(10) $$\rho(2\rho - 1) + \tau - \alpha \leq 1.$$

Next, write $a = |U_1|/n$, where without loss of generality

(11) $$1/4 - b \leq a \leq 1/2.$$

Since $d_{\mathbf{i}}(J) = 1$ for any $\mathbf{i}$ with $|\mathbf{i}| = 2$, by Proposition 8.7(ii) applied to each such $\mathbf{i}$ we have

(12) $$\tau a + (\rho + \alpha)(1 - a) \geq 3/4 - c,$$
(13) $$\tau'(1 - a) + (\rho' + \alpha)a \geq 3/4 - c,$$
(14) $$(\rho + \alpha)a + (\rho' + \alpha)(1 - a) \geq 3/4 - c.$$

We claim that inequalities (9)-(14) imply the estimates
(15) $\quad a = 1/4 \pm \beta, \quad \tau \le \beta, \quad \rho \ge 1 - \beta, \quad \rho' = 2/3 \pm \beta, \quad \tau' = 7/9 \pm \beta.$

The proof of this claim requires some calculations, which are tiresome, but not difficult. We first consider inequality (12). The left hand side is linear in a, so its maximum is attained at (at least) one of the extreme values $a = 1/2$ and $a = 1/4 - b$. Thus we either have (A) $3/4 - c \le \tau/2 + (\rho + \alpha)/2$ or (B) $3/4 - c \le (1/4 - b)\tau + (\rho + \alpha)(3/4 + b)$. We will show that the first case (A) leads to a contradiction. For it implies

$$1 - 2c = 2(3/4 - c) - 1/2 \le \tau + \rho + \alpha - 1/2 = \tau + \rho(2\rho - 1) - (2\rho - 1)^2/2 + \alpha$$
$$\le 1 - (2\rho - 1)^2/2 + 2\alpha$$

by (10), so $(2\rho - 1)^2 \le 5\alpha$. Then $\rho = 1/2 \pm \sqrt{5\alpha}$, so (12) gives

$$3/4 - c \le a + (1/2 + \sqrt{6\alpha})(1 - a) = 1/2 + \sqrt{6\alpha} + (1/2 - \sqrt{6\alpha})a,$$

so $a \ge 1/2 - 9\sqrt{\alpha}$. Since $a \le 1/2$, we can write $a = 1/2 \pm 9\sqrt{\alpha}$, and apply the same reasoning to inequality (13). This gives $3/4 - c \le (\tau' + \rho' + \alpha)(1/2 + 9\sqrt{\alpha})$, so

$$1 - 2c \le (\tau' + \rho' + \alpha)(1 + 18\sqrt{\alpha}) - 1/2 \le \tau' + \rho'(2\rho' - 1) - (2\rho' - 1)^2/2 + 38\sqrt{\alpha}$$
$$\le 1 - (2\rho' - 1)^2/2 + 39\sqrt{\alpha},$$

by (9). Then $(2\rho' - 1)^2 \le 80\sqrt{\alpha}$, so $\rho' = 1/2 \pm 10\alpha^{1/4}$. But now substituting a, ρ, ρ' in (14) gives a contradiction. Thus case (A) is impossible, so we must have case (B). This implies

$$1 - 4c = 4(3/4 - c) - 2 \le \tau + 3\rho + 4\alpha - 2 = \tau + \rho(2\rho - 1) - 2(\rho - 1)^2 + 4\alpha$$
$$\le 1 - 2(\rho - 1)^2 + 5\alpha$$

by (10). Then $(\rho - 1)^2 \le 3\alpha$, so $\rho \ge 1 - \sqrt{3\alpha}$. Then (10) gives

$$\tau \le 1 + \alpha - (1 - \sqrt{3\alpha})(1 - 2\sqrt{3\alpha}) < 6\sqrt{\alpha}.$$

Now (12) gives $3/4 - c \le 6\sqrt{\alpha} + (1 + \alpha)(1 - a)$, so

$$a < 1/4 + 7\sqrt{\alpha}.$$

Then (14) gives $3/4 - c \le (1 + \alpha)(1/4 + 7\sqrt{\alpha}) + (\rho' + \alpha)(3/4 + b)$, so

$$\rho' \ge 2/3 - 10\sqrt{\alpha}.$$

Now (9) implies

$$\tau' \le 1 + \alpha - (2/3 - 10\sqrt{\alpha})(1/3 - 20\sqrt{\alpha}) < 7/9 + 20\sqrt{\alpha}.$$

Next, since $\tau' \le 1 + \alpha - \rho'(2\rho' - 1)$ by (9), substituting in (13) gives

$$3/4 - c \le (1 + \alpha - \rho'(2\rho' - 1))(3/4 + b) + (\rho' + \alpha)(1/4 + 7\sqrt{\alpha})$$
$$\le 3/4 - \frac{3}{2}\rho'(\rho' - 2/3) + 8\sqrt{\alpha},$$

so $\rho' \le 2/3 + 9\sqrt{\alpha}$. Finally, substituting this in (13) gives

$$3/4 - c \le \tau'(3/4 + b) + (2/3 + 10\sqrt{\alpha})(1/4 + 7\sqrt{\alpha}),$$

so $\tau' \ge 7/9 - 10\sqrt{\alpha}$. Thus we have verified all the estimates in (15).

Applying (15) to the vertex $v \in U_1' \setminus Z$ chosen above, we see that the graph $G(v)[U_2]$ has edge density at least $2/3 - \beta$ and triangle density at most $2/9 + 2\beta$. So by Theorem 8.11(ii) we may partition U_2 into sets V_2, V_3 and V_4 each of size at least $n/4 - \gamma n$ such that for each $i \in \{2,3,4\}$ we have $d(G(v)[V_i]) \leq \beta'$. By Theorem 8.11(i), the triangle density of $G(v)[U_2]$ is at least $\rho'(2\rho' - 1) \geq 2/9 - 2\beta$. Since $d(G[U_2]) = \tau' \geq 7/9 - \beta$, we see that all but at most $4\beta\binom{|U_2|}{3}$ triples in U_2 either form triangles in $G(v)[U_2]$ or edges in $G[U_2]$; otherwise, we would have at least $\beta\binom{|U_2|}{3} > \sqrt{b}n^3$ odd tetrahedra containing v, contradicting $v \notin Z$. Now write $V_1 := U_1$ and consider index vectors with respect to the partition (V_1, V_2, V_3, V_4). Then for each $i \in \{2,3,4\}$, since $d(G(v)[V_i]) \leq \beta'$ we have $d_{3\mathbf{u}_i}(J) = d(G[V_i]) \geq 1 - \gamma$. Also, since all but at most $\beta'n^3$ triples of index $(0,1,1,1)$ are triangles of $G(v)$ we have $d_{(0,1,1,1)}(J) < 100\beta'$. Applying Proposition 8.7(ii) to $\mathbf{i} = (2,0,0,0)$ we have

$$d_{(3,0,0,0)}(J)|V_1| + d_{(2,1,0,0)}(J)|V_2| + d_{(2,0,1,0)}(J)|V_3| + d_{(2,0,0,1)}(J)|V_4| \geq (3/4 - c)n.$$

Since $d_{(3,0,0,0)}(J) = d(G[V_1]) = \tau \leq \beta$ and $|V_1| = (1/4 \pm \beta)n$, each of $d_{(2,1,0,0)}(J)$, $d_{(2,0,1,0)}(J)$, $d_{(2,0,0,1)}(J)$ must be at least $1-\gamma$. Similarly, since $d_{(0,1,1,1)}(J) < 100\beta'$, applying Proposition 8.7(ii) to each of $(0,1,1,0)$, $(0,1,0,1)$ and $(0,0,1,1)$ we see that $d_\mathbf{i}(J) \geq 1 - \gamma$ for all remaining good index vectors. □

8.4. Divisibility barriers with more parts

Having considered divisibility barriers with two parts, we now consider the possibility of divisibility barriers with three or more parts. Indeed, in this section we show that, if there are no divisibility barriers with two parts, then there are no divisibility barriers with more than two parts. We consider the cases of three parts and four parts separately (there cannot be more because of the minimum size of the parts). We will repeatedly use the following observation, which is immediate from Proposition 8.7(ii).

PROPOSITION 8.13. *Let $\mu \ll d$ and $\mathcal{P}$ partition a set X of n vertices into r parts. Suppose J is a k-complex on X with $d_\mathbf{i}(J) \geq d$ and $\delta_{|\mathbf{i}|}(J) \geq dn$ for some index vector $\mathbf{i} \in \mathbb{Z}^r$. Then there is some $j \in [r]$ such that $\mathbf{i} + \mathbf{u}_j \in L^\mu_\mathcal{P}(J_{|\mathbf{i}|+1})$, i.e. there are at least $\mu n^{|\mathbf{i}|+1}$ edges in J with index vector $\mathbf{i} + \mathbf{u}_j$.*

We start by considering divisibility barriers with three parts. Recall that a lattice $L \subseteq \mathbb{Z}^d$ is transferral-free if it does not contain any difference $\mathbf{u}_i - \mathbf{u}_j$ of distinct unit vectors $\mathbf{u}_i, \mathbf{u}_j$.

LEMMA 8.14. *Let $1/n \ll c \ll \mu \ll 1$, and G be a 3-graph on n vertices with $\delta(G) \geq (3/4 - c)n$. Write $J = J_4(G)$ for the clique 4-complex of G. Then there is no partition $\mathcal{P}$ of $V(G)$ into three parts of size at least $n/4 - \mu n$ such that $L^\mu_\mathcal{P}(J_4)$ is transferral-free and incomplete.*

PROOF. Introduce constants α, β, β' with $\mu \ll \alpha \ll \beta \ll \beta' \ll 1$. We need the following claim.

CLAIM 8.15. *Suppose (U_1, U_2) is a partition of V with $|U_1| = an$ and $1/4 - \alpha \leq a \leq 1/2 + \alpha$. Then at least one of the following holds:*

(i) there are at least μn^4 tetrahedra with 3 vertices in U_1,
(ii) $a = 1/4 \pm \beta$ and $d_{(2,1)}(J) > 1 - \beta'$,
(iii) $a = 1/2 \pm \beta$ and $d_{(1,2)}(J) > 1 - \beta'$.

To prove the claim, suppose option (i) does not hold, i.e. there are at most μn^4 tetrahedra with 3 vertices in U_1. We repeat the first part of the proof of Lemma 8.12

with μ in place of b. Define τ, τ', ρ, ρ' as in that proof. Then there is a vertex $v \in U_2$ with $d(G(v)[U_1]) \geq \rho$ that belongs to at most $\sqrt{\mu}n^3$ odd tetrahedra (interchanging the roles of U_1 and U_2 from before). It follows that $G(v)[U_1]$ has triangle density at most $1 - \tau + \alpha$, so (10) holds. Also (12) and (14) hold as before.

First we show that $1/4 + \beta < a < 1/2 - \beta$ leads to a contradiction. Since (12) is linear in a, we either have (A) $3/4 - c \leq (1/4 + \beta)\tau + (3/4 - \beta)(\rho + \alpha)$ or (B) $3/4 - c \leq (1/2 - \beta)\tau + (1/2 + \beta)(\rho + \alpha)$. Consider case (A). It implies

$$1 - 4c = 4(3/4 - c) - 2 \leq \tau + 3\rho - 2 - 4\beta(\rho - \tau - \alpha/\beta) - \alpha$$
$$= \tau - \alpha + \rho(2\rho - 1) - 2(\rho - 1)^2 - 4\beta(\rho - \tau - \alpha/\beta)$$
$$\leq 1 - 2(\rho - 1)^2 - 4\beta(\rho - \tau - \alpha/\beta),$$

so $2(\rho - 1)^2 \leq 4c - 4\beta(\rho - \tau - \alpha/\beta)$. However, this implies $\rho > 1 - 2\sqrt{\beta}$, so $\tau < 10\sqrt{\beta}$ by (10), which contradicts $4c - 4\beta(\rho - \tau - \alpha/\beta) \geq 0$.

Now consider case (B). It implies

$$1 - 2c = 2(3/4 - c) - 1/2 \leq \tau + \rho - 1/2 - 2\beta(\tau - \rho - \alpha/\beta)$$
$$= \tau + \rho(2\rho - 1) - (2\rho - 1)^2/2 - 2\beta(\tau - \rho - \alpha/\beta)$$
$$\leq 1 + \alpha - (2\rho - 1)^2/2 - 2\beta(\tau - \rho - \alpha/\beta),$$

so $(2\rho - 1)^2 + 4\beta(\tau - \rho - \alpha/\beta) \leq 4c + 2\alpha$. Then $\rho = 1/2 \pm 3\sqrt{\beta}$ and $\beta(\tau - \rho - \alpha/\beta) \leq c + \alpha/2$. Now (12) gives $3/4 - c \leq \tau(1/2 - \beta) + (1/2 + 3\sqrt{\beta} + \alpha)(3/4 - \beta)$, so $\tau \geq 3/4$. But now $\beta/5 \leq \beta(\tau - \rho - \alpha/\beta) \leq c + \alpha/2$, which is a contradiction.

Next suppose that $a = 1/4 \pm \beta$. By (12) we have $3/4 - c \leq (1/4 + \beta)\tau + (3/4 + \beta)(\rho + \alpha)$. Similarly to case (A) we have

$$1 - 4c = 4(3/4 - c) - 2 \leq \tau + 3\rho - 2 + 9\beta \leq 1 - 2(\rho - 1)^2 + \alpha + 9\beta,$$

so $d_{(2,1)}(J) = \rho > 1 - \beta'$, which is option (ii). Finally, suppose that $a = 1/2 \pm \beta$. By (12) we have $3/4 - c \leq (1/2 + \beta)(\tau + \rho + \alpha)$. Similarly to case (B) we have

$$1 - 2c = 2(3/4 - c) - 1/2 \leq \tau + \rho - 1/2 + 5\beta \leq 1 + \alpha - (2\rho - 1)^2/2 + 5\beta,$$

so $\rho = 1/2 \pm 2\sqrt{\beta}$. Now (14) gives $3/4 - c \leq (1/2 + \beta)(\rho + \alpha + \rho' + \alpha)$, so $d_{(1,2)}(J) = \rho' > 1 - \beta'$, which is option (iii). This proves Claim 8.15.

Returning to the proof of the lemma, suppose for a contradiction we have a partition $\mathcal{P}$ of V into parts V_1, V_2, V_3 of size at least $n/4 - \mu n$ such that $L_\mathcal{P}^\mu(J_4)$ is transferral-free and incomplete. Without loss of generality $|V_1| \geq n/3$. Also, since each of V_1, V_2, V_3 has size at least $n/4 - \mu n$, each has size at most $n/2 + 2\mu n$. Now recall that $\delta(J) \geq (n, n-1, (3/4 - c)n, (1/4 - 3c)n)$ by (8). Since J_2 is complete, for any $\mathbf{i}$ with $|\mathbf{i}| = 2$, Proposition 8.7(ii) gives

$$(3/4 - c)n \leq \sum_{j \in [3]} d_{\mathbf{i}+\mathbf{u}_j}(J)|V_j| \leq d_{\mathbf{i}+\mathbf{u}_1}(J)|V_1| + n - |V_1|.$$

This gives

$$(1/4 + c)n \geq (1 - d_{\mathbf{i}+\mathbf{u}_1}(J))|V_1| \geq (1 - d_{\mathbf{i}+\mathbf{u}_1}(J))n/3,$$

so $d_{\mathbf{i}+\mathbf{u}_1}(J) \geq 1/5$. Thus $d_\mathbf{i}(J) \geq 1/5$ for every $\mathbf{i} = (i_1, i_2, i_3)$ with $|\mathbf{i}| = 3$ and $i_1 \geq 1$. For each such $\mathbf{i}$, Proposition 8.13 gives some $j \in [3]$ with $\mathbf{i} + \mathbf{u}_j \in L := L_\mathcal{P}^\mu(J_4)$. In particular, there is some $j \in [3]$ such that $(1,1,1) + \mathbf{u}_j \in L$. Without loss of generality $j = 1$ or $j = 2$, as if $j = 3$ we can rename V_2 and V_3 to get $j = 2$.

Suppose first that $j = 1$, i.e. $(2,1,1) \in L$. We apply Claim 8.15 to the partition with $U_1 = V_1$ and $U_2 = V_2 \cup V_3$. Option (i) cannot hold, as then either $(3,1,0) \in L$, so $(3,1,0) - (2,1,1) = (1,0,-1) \in L$, or $(3,0,1) \in L$, so $(3,0,1) - (2,1,1) = (1,-1,0) \in L$, which contradicts the fact that L is transferral-free. Also, option (ii) cannot hold, since $|V_1| \geq n/3$. Thus option (iii) holds, i.e. $|U_1| = an$ with $a = 1/2 \pm \beta$ and $d_{(1,2)}(J) > 1 - \beta'$ with respect to (U_1, U_2). It follows that $d_{\mathbf{i}}(J) > 1 - 10\beta'$ with respect to $\mathcal{P}$ for any $\mathbf{i} = (i_1, i_2, i_3)$ with $i_1 = 1$, $i_2 + i_3 = 2$ and $i_2, i_3 \geq 0$. Now Proposition 8.7(ii) applied to $(0,1,1)$ gives $(3/4 - c)n \leq (1/2 + \beta)n + (d_{(0,2,1)}(J) + d_{(0,1,2)}(J))(1/4 + 2\beta)n$, so without loss of generality $d_{(0,2,1)}(J) > 1/3$. Since $J = J_4(G)$ is a clique complex, we have $J_{(1,2,1)} = J^*_{(1,2,1)}$, so we can apply Proposition 8.7(iii) to get $d_{(1,2,1)}(J) \geq 1 - 30\beta' - 2/3 \geq 1/4$. Thus $(1,2,1) \in L$, so $(1,2,1) - (2,1,1) = (-1,1,0) \in L$, again contradicting the fact that L is transferral-free.

Now suppose that $j = 2$, i.e. $(1,2,1) \in L$. We apply Claim 8.15 to the partition with $U_1 = V_2$ and $U_2 = V_1 \cup V_3$. As in the case $j = 1$, option (i) cannot hold since L is transferral-free. Also, option (iii) cannot hold, since $|V_1| \geq n/3$ and $|V_3| \geq n/4 - \mu n$. Thus option (ii) holds, i.e. $|U_2| = an$ with $a = 1/4 \pm \beta$ and $d_{(2,1)}(J) > 1 - \beta'$ with respect to (U_1, U_2). It follows that $d_{(1,2,0)}(J)$ and $d_{(0,2,1)}(J)$ are both at least $1 - 10\beta'$ (say) with respect to $\mathcal{P}$. Next note that $(2,2,0) \notin L$ and $(2,1,1) \notin L$ since L is transferral-free. Thus $d_{(2,2,0)}(J)$ and $d_{(2,1,1)}(J)$ are both at most 500μ (say). Applying Proposition 8.7(ii) to $(2,1,0)$ gives $(1/4 - 3c)n \leq d_{(3,1,0)}(J)|V_1| + 1000\mu n$, so $d_{(3,1,0)}(J) \geq 1/3$ (say). Then Proposition 8.7(i) gives $d_{(2,1,0)}(J) \geq 1/3$. Now choose $e = xx'yy'$ in $K(V)_{(2,2,0)}$ uniformly at random, with $x, x' \in V_1$ and $y, y' \in V_2$. Given xx', let $\rho_{xx'}|V_2|$ be the number of $v_2 \in V_2$ such that $xx'v_2 \in G$. Then $\mathbb{E}_{xx'} \rho_{xx'} = d_{(2,1,0)}(J) \geq 1/3$. So the probability that $xx'y$ and $xx'y'$ are both edges is at least $\mathbb{E}_{xx'} \rho_{xx'}(\rho_{xx'} - 5/n) \geq 1/10$, by Cauchy-Schwartz. Also, each of xyy' and $x'yy'$ is an edge with probability at least $d_{(1,2,0)}(J) \geq 1 - 10\beta'$. Since $J = J_4(G)$ is a clique complex, we deduce $d_{(2,2,0)}(J) = \mathbb{P}(e \in J_{(2,2,0)}) > 1/11$. But this contradicts $(2,2,0) \notin L$.

In either case we obtain a contradiction to the existence of the divisibility barrier $\mathcal{P}$. □

Now we consider divisibility barriers with four parts.

LEMMA 8.16. *Let $1/n \ll c \ll \mu \ll 1$, and G be a 3-graph on n vertices with $\delta(G) \geq (3/4 - c)n$. Write $J = J_4(G)$ for the clique 4-complex of G. Then there is no partition $\mathcal{P}$ of $V(G)$ into four parts of size at least $n/4 - \mu n$ such that $L^\mu_\mathcal{P}(J_4)$ is transferral-free and incomplete.*

PROOF. We introduce constants μ', β, β' with $\mu \ll \mu' \ll \beta \ll \beta' \ll 1$. Suppose for a contradiction we have a partition $\mathcal{P}$ of V into parts V_1, V_2, V_3, V_4 of size at least $n/4 - \mu n$ such that $L = L^\mu_\mathcal{P}(J_4)$ is incomplete and transferral-free. Note that all parts have size at most $n/4 + 3\mu n$. Recall that $\delta(J) \geq (n, n-1, (3/4-c)n, (1/4-3c)n)$ by (8). Now we need the following claim.

CLAIM 8.17.
(i) *If $\mathbf{i} \in \mathbb{Z}^4$ with $|\mathbf{i}| = 4$ and $d_{\mathbf{i}}(J) \geq \beta$ then $d_{\mathbf{i}}(J) \geq 1 - \beta$.*
(ii) *If $\mathbf{i} \in \mathbb{Z}^4$ with $|\mathbf{i}| = 3$ and $d_{\mathbf{i}}(J) \geq 2\beta$ then $d_{\mathbf{i}}(J) \geq 1 - \beta$.*

To prove the claim, first consider any $\mathbf{i} \in \mathbb{Z}^4$ with $|\mathbf{i}| = 4$ and $d_{\mathbf{i}}(J) \geq \beta$. Note that $\mathbf{i} \in L$. By Proposition 8.7(iv) there is some $\mathbf{i}' = \mathbf{i} - \mathbf{u}_j \in \partial \mathbf{i}$ such that

$d_{\mathbf{i}'}(J) \geq d_{\mathbf{i}}(J)^{3/4} + O(1/n)$. Let B be the set of edges $e' \in J_{\mathbf{i}'}$ that lie in at least $\mu'n$ edges $e \in J_4$ with $\mathbf{i}(e) \neq \mathbf{i}$. Then there is some $j' \neq j$ for which the number of edges with index vector $\mathbf{i} - \mathbf{u}_j + \mathbf{u}_{j'}$ is at least $|B|\mu'n/12$. This must be less than μn^4, otherwise we have $\mathbf{i} - \mathbf{u}_j + \mathbf{u}_{j'} \in L$, so $\mathbf{u}_j - \mathbf{u}_{j'} \in L$, contradicting the fact that L is transferral-free. It follows that $d_{\mathbf{i}'}(B) \leq \mu'$ (say), so $d_{\mathbf{i}'}(J \setminus B) \geq d_{\mathbf{i}}(J)^{3/4} - 2\mu'$. Next we show that if $e' \in J \setminus B$ then $e' \cup \{x\} \in J_{\mathbf{i}}$ for all but $2\mu'n$ vertices $x \in V_j$. This holds because there are at least $\delta_3(J) \geq (1/4 - 3c)n$ vertices x such that $e' \cup \{x\} \in J_4$, all but at most $\mu'n$ of these lie in V_j (as $e' \notin B$), and all parts have size at most $n/4 + 3\mu n$. It follows that

$$d_{\mathbf{i}}(J) \geq (1 - 10\mu')d_{\mathbf{i}'}(J \setminus B) \geq (1 - 10\mu')(d_{\mathbf{i}}(J)^{3/4} - 2\mu').$$

Since $d_{\mathbf{i}}(J) \geq \beta$ this implies $d_{\mathbf{i}}(J) \geq 1 - \beta$, so we have proved (i). For (ii), observe that by Proposition 8.7(ii) there must be some $\mathbf{i}'$ such that $|\mathbf{i}'| = 4$, $\mathbf{i} \leq \mathbf{i}'$ and $d_{\mathbf{i}'}(J) \geq (3/4 - c)2\beta \geq \beta$. Then by (i) we have $d_{\mathbf{i}}(J) \geq 1 - \beta$, so by Proposition 8.7(i) we have $d_{\mathbf{i}'}(J) \geq d_{\mathbf{i}}(J) \geq 1 - \beta$. This completes the proof of Claim 8.17.

Returning to the proof of the lemma, we show next that $d_{\mathbf{i}}(J) \leq 2\beta$ whenever $\mathbf{i}$ is a permutation of $(3,0,0,0)$. For suppose this fails, say for $\mathbf{i} = (3,0,0,0)$. Then Claim 8.17 gives $d_{(3,0,0,0)}(J) \geq 1 - \beta$. Now Proposition 8.7(iii) gives $d_{(4,0,0,0)}(J) \geq 1 - 4\beta$, so $(4,0,0,0) \in L$. Next, Proposition 8.7(ii) for $\mathbf{i}' = (2,0,0,0)$ gives $(3/4 - c)n \leq (n/4 + 3\mu n)\sum_{j \in [4]} d_{\mathbf{i}' + \mathbf{u}_j}(J)$, so without loss of generality $d_{(2,1,0,0)}(J) \geq 2\beta$. Then Claim 8.17 gives $d_{(2,1,0,0)}(J) \geq 1 - \beta$. Now Proposition 8.7(iii) gives $d_{(3,1,0,0)}(J) \geq 1 - 4\beta$, so $(3,1,0,0) \in L$. But then $(4,0,0,0) - (3,1,0,0) = (1,-1,0,0) \in L$ contradicts the fact that L is transferral-free. Thus $d_{\mathbf{i}}(J) \leq 2\beta$ whenever $\mathbf{i}$ is a permutation of $(3,0,0,0)$.

Now returning to the above inequality

$$(3/4 - c)n \leq (n/4 + 3\mu n)\sum_{j \in [4]} d_{\mathbf{i}' + \mathbf{u}_j}(J),$$

where $\mathbf{i}'$ is any permutation of $(2,0,0,0)$, we see that for any $\mathbf{i}''$ which is a permutation of $(2,1,0,0)$ we have $d_{\mathbf{i}''}(J) \geq 2\beta$, and therefore $d_{\mathbf{i}''}(J) \geq (1 - \beta)$ by Claim 8.17. Also, applying Proposition 8.7(ii) to $(1,1,0,0)$, we see that without loss of generality $d_{(1,1,1,0)}(J) \geq 2\beta$. Then Claim 8.17 gives $d_{(1,1,1,0)}(J) \geq 1 - \beta$. Now each of $d_{(2,2,0,0)}(J)$ and $d_{(2,1,1,0)}(J)$ is at least $1 - 4\beta$ by Proposition 8.7(iii). It follows that L contains $(2,2,0,0)$, $(2,1,1,0)$ and $(2,2,0,0) - (2,1,1,0) = (0,1,-1,0)$, again contradicting the fact that L is transferral-free. Thus there is no such partition $\mathcal{P}$. □

8.5. The main case of Theorem 1.1

As outlined earlier, Theorem 2.9 will imply the desired result, except for 3-graphs that are close to the extremal configuration. We start with a lemma that handles such 3-graphs. The proof requires a bound on the minimum vertex degree that guarantees a perfect matching in a 4-graph. We just quote the following (slight weakening of a) result of Daykin and Häggkvist [10] that suffices for our purposes. (The tight bound was recently obtained by Khan [28].)

THEOREM 8.18. *Let H be a 4-graph on n vertices, where $4 \mid n$, such that every vertex belongs to at least $\frac{3}{4}\binom{n}{3}$ edges. Then H contains a perfect matching.*

The following lemma will be used for 3-graphs that are close to the extremal configuration. Note that we will be able to satisfy the condition that $|V_1|$ is even by applying Lemma 8.10.

LEMMA 8.19. *Suppose that $1/n \ll \gamma \ll c \ll 1$. Let V be a set of $4n$ vertices partitioned into V_1, V_2, V_3, V_4 each of size at least $n - \gamma n$. Suppose J is a 4-complex on V such that*

 (i) $d(J_4[V_j]) \geq 1 - \gamma$ for each $j \in \{2, 3, 4\}$,
 (ii) *for each* $\mathbf{i} \in (2, 1, 1, 0), (2, 1, 0, 1), (2, 0, 1, 1)$ *we have* $d_\mathbf{i}(J) \geq 1 - \gamma$,
 (iii) *for every vertex $v \in V$ at least cn^4 edges of J_4 contain v and have an even number of vertices in V_1, and*
 (iv) $|V_1|$ *is even.*

Then J_4 contains a perfect matching.

PROOF. We say that an edge of J_4 is *even* if it has an even number of vertices in V_1, and *odd* otherwise. For each $j \in \{2, 3, 4\}$, we say that a vertex $v \in V_j$ is *good* if there are at least $(1 - 2\sqrt{\gamma})\binom{n}{3}$ edges of $J_4[V_j]$ which contain v. By (i) at most $\sqrt{\gamma} n$ vertices of each of V_2, V_3 and V_4 are bad; here we note that each part has size at most $n + 3\gamma n$. We say that a pair $u, v \in V_1$ is *good* if for each $\mathbf{i} \in \{(2, 0, 1, 1), (2, 1, 0, 1), (2, 1, 1, 0)\}$ there are at least $(1 - 2\sqrt{\gamma})n^2$ edges $e \in J_4$ with $u, v \in e$ and $\mathbf{i}(e) = \mathbf{i}$. For each such $\mathbf{i}$, by (ii) at most $\sqrt{\gamma}\binom{n}{2}$ pairs $u, v \in V_1$ do not have this property, and so at most $3\sqrt{\gamma}\binom{n}{2}$ pairs $u, v \in V_1$ are bad. We say that a vertex $u \in V_1$ is *good* if it lies in at least $(1 - 2\gamma^{1/4})n$ good pairs $u, v \in V_1$. Then at most $3\gamma^{1/4}n$ vertices of V_1 are bad. In total, the number of bad vertices is at most $\gamma^{1/5}n$, say.

Now let E be a maximal matching in J such that every edge in E has an even number of vertices in V_1 and contains a bad vertex. We claim that E covers all bad vertices. For suppose some bad vertex v is not covered by E. Since each edge in E contains a bad vertex, at most $4\gamma^{1/5}n$ vertices are covered by E, so at most $4\gamma^{1/5}n^4$ edges of J_4 contain a vertex covered by E. Then by (iii) we may choose an edge $e \in J_4$ which contains v, has an even number of vertices in V_1 and doesn't contain any vertex covered by E, contradicting maximality of E. So E must cover every bad vertex of J. We will include E in our final matching and so we delete the vertices it covers. We also delete the vertices covered by another matching E' of at most 16 edges, disjoint from E, so as to leave parts V_1', V_2', V_3', V_4' such that 8 divides $|V_1'|$ and 4 divides each of $|V_2'|, |V_3'|, |V_4'|$. The edges in E' will have index $(2, 1, 1, 0), (2, 1, 0, 1)$ or $(2, 0, 1, 1)$. Note that by (ii) we can greedily choose a matching disjoint from E containing 16 edges of each of these indices, so we only need to decide how many of these we want to include in E'. First we arrange that there are an even number of remaining vertices in each part. Note that $|V|$ and $|V_1|$ are even, and the number of vertices remaining in V_1 is still even, as we only used edges with an even number of vertices in V_1. Then an even number of parts have an odd number of vertices remaining (after the deletion of E). If there are two such parts, say V_2 and V_4, then we remedy this by including an edge of index $(2, 1, 0, 1)$ in E', so we can assume there are no such parts. Now there are an even number of parts in which the number of vertices remaining is not divisible by 4. If V_1 is one of these parts we include an edge of each index $(2, 1, 1, 0), (2, 1, 0, 1), (2, 0, 1, 1)$ in E'; then each part has an even number of remaining vertices and the number in V_1 is divisible by 4. There may still be two parts where the number of vertices remaining is not divisible by 4. If so, say they are V_3 and V_4, we include 2 edges of index $(2,0,1,1)$ in E'. Thus we arrange that the number of vertices remaining in each part is divisible by 4. Finally, if the number of remaining vertices in V_1 is not divisible by 8 then we include 2 more edges of each index in E'. Thus we obtain E' with the desired properties. We delete the vertices of $E \cup E'$ and let V_1', V_2', V_3', V_4' denote the remaining parts.

Next we choose a perfect matching M_1 of good pairs in V_1'. This is possible because $|V_1'|$ is even, and any vertex in V_1' is good, in that it was incident to at least $(1 - 2\gamma^{1/4})n$ good pairs in V_1, so is still incident to at least $n/2$ good pairs in V_1'. Now we greedily construct a matching in J_4, by considering each pair in M_1 in turn and choosing an edge of index $(2,1,1,0)$ containing that pair. There are at most $n/2$ pairs in M_1, so when we consider any pair uv in M_1, at least $n/3$ vertices in each of V_2' and V_3' are still available, in that they have not already been selected when we considered some previous pair in M_1. Since uv is good, at most $3\sqrt{\gamma}n^2 < (n/3)^2$ of these pairs do not form an edge with uv, so we can form an edge as required. Thus we construct a matching E'' that covers V_1'. We delete the vertices of E'' and let V_2'', V_3'', V_4'' denote the remaining parts. Note that the number of vertices used by E'' in each of V_2' and V_3' is $|V_1'|/2$, which is divisible by 4. Since 4 divides each of $|V_2'|$, $|V_3'|$, $|V_4'|$, it also divides each of $|V_2''|$, $|V_3''|$, $|V_4''|$. Furthermore, for any $j \in \{2,3,4\}$ and $x \in V_j$, since x is good, the number of edges of $J_4[V_j'']$ containing x is at least $\binom{|V_j''|}{3} - 3\sqrt{\gamma}\binom{n}{3} > \frac{3}{4}\binom{|V_j''|}{3}$, so $J_4[V_j'']$ contains a perfect matching by Theorem 8.18. Combining these matchings with E, E' and E'' we obtain a perfect matching in J_4. □

We can now give the proof of Theorem 1.1, as outlined at the start of this chapter. Suppose that G is a 3-graph on n vertices, where n is sufficiently large and divisible by 4. For now we assume $\delta(G) \geq 3n/4 - 1$, postponing the case that $8 \mid n$ and $\delta(G) = 3n/4 - 2$ to the final section. We introduce constants α, β with $1/n \ll \alpha \ll \beta \ll 1$. We will show that G has a perfect tetrahedron packing; equivalently, that J_4 has a perfect matching, where $J = J_4(G)$ is the clique 4-complex of G. By (8) we have $\delta(J) \geq (n, n-1, 3n/4 - 2, n/4 - 3)$, so we can apply Theorem 2.9. Supposing that J_4 does not contain a perfect matching, we conclude that there is a space barrier or a divisibility barrier. As in the proof of Theorem 8.2 there are no space barriers by Propositions 8.3 and 8.6, so we may choose a minimal divisibility barrier, i.e. a partition $\mathcal{Q}$ of $V(G)$ into parts of size at most $n/4 - \mu n$ such that $L_\mathcal{Q}^\mu(J_4)$ is transferral-free and incomplete. We have excluded all but one possibility for $\mathcal{Q}$. Indeed, $\mathcal{Q}$ cannot have more than two parts by Lemmas 8.14 and 8.16, so must have two parts. By Lemmas 8.8 and 8.9, we deduce that $\mathcal{Q}$ partitions $V(G)$ into parts U and V of size at least $n/4 - \alpha n$ such that at most αn^4 edges of J_4 have an odd number of vertices in each part.

By Lemma 8.12, there is a partition $\mathcal{P}$ of $V(G)$ into parts V_1, V_2, V_3 and V_4 each of size at least $n/4 - \beta n$, so that $V_1 = U$ (without loss of generality) and $d_\mathbf{i}(G) \geq 1 - \beta$ for each good $\mathbf{i}$ (recall that good index vectors are those that appear in the extremal example). To apply Lemma 8.19 we also need to arrange that $|V_1|$ is even, and that every vertex is in many even edges of J_4, where we say that an edge of J_4 is odd or even according to the parity of its intersection with V_1. We say that a vertex is *good* if it lies in fewer than $n^3/400$ odd edges of J_4 (and *bad* otherwise). Then at most $1600\alpha n$ vertices are bad, since J_4 has at most αn^4 odd edges. We let V_1' consist of all good vertices in V_1 and all bad vertices in $V_2 \cup V_3 \cup V_4$, let V_2' consist of all good vertices in V_2 and all bad vertices in V_1, and let V_3' and V_4' consist of all good vertices in V_3 and V_4 respectively. Note that any vertex that has a different index with respect to the partition $\mathcal{P}'$ of $V(G)$ into the V_j' than with respect to $\mathcal{P}$ is bad, so we have $|V_1'| = n/4 \pm 2\beta n$. So by Lemma 8.10, there exists a copy of K_4^3 in G (i.e. an edge of J_4) with an odd number of vertices in V_1'. If $|V_1'|$ is odd, we choose one such edge of J_4, and delete the vertices of this edge from V_1', V_2', V_3', V_4', and J. Thus we can arrange that $|V_1'|$ is even.

Having changed the index of at most $1600\alpha n$ vertices of J and deleted at most four vertices, we have $|V_i'| \geq n/4 - 2\beta n$ for each $i \in [4]$, and $d_{\mathbf{i}}(G) \geq 1 - 2\beta$ for each good $\mathbf{i}$ with respect to the partition into V_1', V_2', V_3', V_4'. It remains to show that any undeleted vertex v belongs to many edges with an even number of vertices in V_1'. To avoid confusion, we now use the terms X-odd and X-even, where X is V_1 or V_1', to describe edges according to the parity of their intersection with X. First suppose that v is good. Then any edge of J_4 containing v that is V_1'-odd but not V_1-odd must contain a bad vertex, so there are at most $1600\alpha n^3$ such edges. Since v belongs to at most $n^3/400$ V_1-odd edges of J_4, it belongs to at most $n^3/400 + 1600\alpha n^3$ V_1'-odd edges of J_4. Also, v belongs to at least $n^3/200$ edges of J_4, by the lower bound on $\delta(J)$, so v belongs to at least $n^3/500$ V_1'-even edges. Now suppose that v is bad. Then any V_1-odd edge of J_4 containing v and no other bad or deleted vertices is V_1'-even. The number of such edges is at least $n^3/400 - 1600\alpha n^3 - 4n^2 \geq n^3/500$. This shows that every vertex is contained in at least $n^3/500$ V_1'-even edges of J_4.

Finally, Lemma 8.19 implies that J_4 has a perfect matching.

8.6. The case when 8 divides n

It remains to consider the case when $8 \mid n$ and $\delta(G) = 3n/4 - 2$. We apply the same proof as in the previous case, noting that we only needed $\delta(G) \geq 3n/4 - 1$ if $|V_1'|$ was odd, when we used it to find a V_1'-odd tetrahedron. Examining the proof of Proposition 8.10, we see that a V_1'-odd tetrahedron exists under the assumption $\delta(G) = 3n/4 - 2$, unless we have $n/4 - 2 \leq |V_1'| \leq n/4$. Furthermore, we only need a V_1'-odd tetrahedron when $|V_1'|$ is odd, so we only need to consider the exceptional case that $|V_1'| = n/4 - 1$ and there is no V_1'-odd tetrahedron.

Let c_1 satisfy $\beta \ll c_1 \ll 1$. Recall that $d_{\mathbf{i}}(G) \geq 1 - 2\beta$ for all good $\mathbf{i}$. Note that for each bad triple e there is a part V_j' such that for every $x \in V_j' \setminus e$, $e \cup \{x\}$ is V_1'-odd, and every triple in $e \cup \{x\}$ apart from e is good. (We describe triples as good or bad according to their index vectors.) If e is an edge, then some such triple is not an edge. Since there is no V_1'-odd tetrahedron, there are at most βn^3 bad edges by Proposition 8.7(iii). For any $2/n < c < 1/4$, we say that a pair is c-bad if it is contained in at most $(3/4 - 2c)n$ good edges. Since $\delta(G) = 3n/4 - 2$, any c-bad pair is contained in at least cn bad edges. If $\beta \ll c$ then there are at most cn^2 c-bad pairs. Since there is no V_1'-odd tetrahedron, any bad edge of G contains at least one $1/30$-bad pair.

Without loss of generality we have $|V_4'| \geq n/4 + 1$. Thus for each $v_2 \in V_2'$ and $v_3 \in V_3'$ there is some $v_4 \in V_4'$ with $v_2 v_3 v_4 \in G$. Each such edge is bad, so contains a $1/30$-bad pair. The bad pair can only be $v_2 v_3$ for at most $c_1 n^2$ such triples. Thus without loss of generality there is a vertex $v \in V_4$ such that at least $n/10$ pairs $v_3 v$ with $v_3 \in V_3'$ are $1/30$-bad. For each such pair there are at least $n/30$ vertices $v_2 \in V_2'$ such that $v_2 v_3 v$ is a bad edge. For each such bad edge and each $v_1 \in V_1'$, some triple in $v_1v_2v_3v$ is not an edge. Since $d_{(1,1,1,0)}(G) \geq 1 - 2\beta$, there are at most $c_1 n$ vertices $v_1 \in V_1'$ such that $v_1v_2v_3$ is not an edge for at least $c_1 n^2$ pairs v_2v_3. Thus for all but at most $c_1 n$ vertices $v_1 \in V_1'$ there are at least $n^2/300 - c_1 n^2$ pairs v_2v_3 such that one of v_1v_2v or v_1v_3v is not an edge. Then v_1v is $1/200$-bad, so there are at least $n/200$ vertices $v_4 \in V_4'$ for which v_1vv_4 is a (bad) edge.

Now for each $v_1 \in V_1'$, $v_4 \in V_4'$ such that v_1vv_4 is an edge and $x \in V_2' \cup V_3'$, at least one of v_1xv_4, v_1xv, xvv_4 is not an edge. Since $d_{(1,1,0,1)}(G)$ and $d_{(1,1,0,1)}(G)$ are at least $1 - 2\beta$, there are at most $c_1 n$ vertices $v_1 \in V_1'$ such that v_1xv_4 is not an edge for at least $c_1 n^2$ pairs xv_4. For $v_1 \in V_1'$, let $N_4(v_1)$ be the set of vertices $v_4 \in V_4'$ such that v_1vv_4 is an edge, and v_1xv_4 is an edge for all but at most $600c_1 n$ vertices $x \in V_2' \cup V_3'$. Let V_1'' be the set of vertices $v_1 \in V_1'$ such that $|N_4(v_1)| \geq n/300$. Then $|V_1' \setminus V_1''| \leq 2c_1 n$, otherwise there are at least $c_1 n$ vertices $v_1 \in V_1' \setminus V_1''$ such that there are at least $n/200 - |N_4(v_1)| \geq n/600$ vertices $v_4 \in V_4'$ such that v_1vv_4 is an edge and $v_4 \notin N_4(v_1)$, so v_1xv_4 is not an edge for at least $600c_1 n \cdot n/600 = c_1 n^2$ pairs xv_4, which is a contradiction.

For each $v_1 \in V_1''$, $v_4 \in N_4(v_1)$ and all but at most $600c_1 n$ vertices $x \in V_2' \cup V_3'$, v_1xv_4 is an edge, so at least one of v_1xv and xvv_4 is not an edge. Since $d(v_1v) + d(vv_4) \geq 2\delta(G) = 3n/2 - 4$, and $|V_j'| \geq n/4 - 2\beta n$ for each $j \in [4]$, precisely one of v_1xv and xvv_4 is not an edge for all but at most $2000c_1 n$ vertices $x \in V_2' \cup V_3'$. Also, v_1xv and xvv_4 are edges for all but at most $2000c_1 n$ vertices $x \in V_1' \cup V_4'$. Now there can be at most $c_1 n$ vertices $v_1 \in V_1''$ such that v_1v_2v is an edge for at least $3000c_1 n$ vertices $v_2 \in V_2'$. Otherwise, since we have at least $n/300$ choices of $v_4 \in N_4(v_1)$, so v_1vv_4 is an edge, and $1000c_1 n$ choices of v_2 such that v_2vv_4 is an edge, there must be at least $c_1^2 n^3$ good triples $v_1v_2v_4$ that are not edges, contradicting $d_{(1,1,0,1)}(G) \geq 1 - 2\beta$. Similarly, there are at most $c_1 n$ vertices $v_1 \in V_1''$ such that v_1v_3v is an edge for at least $3000c_1 n$ vertices $v_3 \in V_3'$. But then we can choose $v_1 \in V_1''$ such that v_1xv is an edge for at most $6000c_1 n$ vertices $x \in V_2' \cup V_3'$, which contradicts $\delta(G) = 3n/4 - 2$.

Thus the exceptional case considered here cannot occur. The rest of the proof goes through as in the previous section, so this proves Theorem 1.1. $\square$

8.7. Strong stability for perfect matchings

We conclude this chapter by applying a similar argument to that of Lemma 8.19 to prove Theorem 1.3. For this we need another theorem of Daykin and Häggkvist [10], which gives a bound on the vertex degree needed to guarantee a perfect matching in a k-partite k-graph.

THEOREM 8.20 ([10]). *Suppose that G is a k-partite k-graph whose vertex classes each have size n, in which every vertex lies in at least $\frac{k-1}{k}n^{k-1}$ edges. Then G contains a perfect matching.*

The proof of Theorem 1.3 begins with Theorem 2.14, from which we obtain a set V_1 whose intersection with almost all edges of G has the same parity. This gives us a great deal of structural information on G, namely that G contains almost every possible edge whose intersection with V_1 has the 'correct' parity. Given a single edge of the opposite parity, we can then delete this edge (if necessary) to 'correct' the parity of $|V_1|$, whereupon the high density of edges of 'correct' parity guarantees the existence of a perfect matching in H by Theorem 8.20. We now give the details of the proof, but first restate the theorem in a slightly different form.

Theorem 1.3. *Suppose that $1/n \ll c \ll 1/k$, $k \geq 3$ and $k \mid n$, and let G be a k-graph on n vertices with $\delta(G) \geq (1/2 - c)n$. Then G does not contain a perfect matching if and only if there is a partition of $V(G)$ into parts V_1, V_2 of size at least $\delta(G)$ and $a \in \{0, 1\}$ so that $|V_1| \neq an/k \mod 2$ and $|e \cap V_1| = a \mod 2$ for all edges e of G.*

PROOF. First observe that if such a partition exists there can be no perfect matching in G. Indeed, the n/k edges in such a matching would each have $|e \cap V_1| = a \mod 2$, giving $|V_1| = an/k \mod 2$, a contradiction. So assume that G contains no perfect matching; to complete the proof it suffices to show that G admits a partition as described.

Introduce new constants b, b' with $1/n \ll b' \ll b \ll c \ll 1/k$. By Theorem 2.14, there is a partition of $V(G)$ into parts V_1, V_2 of size at least $\delta(G)$ and $a \in \{0, 1\}$ so that all but at most $b'n^k$ edges $e \in G$ have $|e \cap V_1| = a \mod 2$. Say that an edge e is *good* if $|e \cap V_1| = a \mod 2$, and *bad* otherwise. We begin by moving any vertex of V_1 which lies in fewer than $n^{k-1}/6(k-1)!$ good edges to V_2, and moving any vertex of V_2 which lies in fewer than $n^{k-1}/6(k-1)!$ good edges to V_1. Having made these moves, we update our definitions of 'good' and 'bad' edges to the new partition (so some edges will have changed from good to bad, and vice versa). We claim that the modified partition has parts of size at least $\delta(G) - bn$, that at most bn^k edges are now bad, and any vertex now lies in at least $n^{k-1}/7(k-1)!$ good edges of G. To see this, first note that the minimum degree of G implies that every vertex of G lies in at least $n^{k-1}/3(k-1)!$ edges of G. So any vertex of G which we moved originally lay in at least $n^{k-1}/6(k-1)!$ bad edges of G; since there were at most $b'n^k$ bad edges in total we conclude that at most $6k!b'n \leq bn$ vertices were moved, giving the bound on the new part sizes. Any edge which is now bad was either one of the at most $b'n^k$ edges which were previously bad, or one of the at most $6k!b'n^k \leq bn^k/2$ edges which contain a vertex we moved, so there are at most bn^k bad edges after the vertex movements. If a vertex was not moved, then it previously lay in at least $n^{k-1}/6(k-1)!$ good edges of G; each of these edges is now good unless it is one of the at most $6k!b'n^{k-1}$ edges which also contain another moved vertex. Similarly, if a vertex was moved, then it previously lay in at least $n^{k-1}/6(k-1)!$ bad edges of G; each of these edges is now good unless it is one of the at most $6k!b'n^{k-1}$ edges which also contain another moved vertex. In either case we conclude that the vertex now lies in at least $n^{k-1}/7(k-1)!$ good edges of G. Having fixed our new partition, we say that a vertex is *bad* if it is contained in at least $k\sqrt{b}n^{k-1}$ bad edges. In particular the set B of bad vertices has size $|B| \leq \sqrt{b}n$.

Suppose first that G contains a bad edge $e^* = \{u_1, \ldots, u_k\}$. Then we may choose vertex-disjoint good edges $e_1, \ldots, e_k$ such that $u_i \in e_i$ and the edges e_i contain no bad vertices except possibly the vertices u_i. Having done this, greedily choose a matching E in G of size at most $|B|$ which covers all bad vertices and is vertex-disjoint from $e_1, \ldots, e_k$ and e^*. Now consider the matchings $E_1 := E \cup \{e_1, \ldots, e_k\}$ and $E_2 := E \cup \{e^*\}$. We will choose either $E^* = E_1$ or $E^* = E_2$, so that, writing V_1' and V_2' for the vertices remaining in V_1 and V_2 respectively after deleting the vertices covered by E^*, and $n' := |V_1' \cup V_2'|$, we have that $|V_1'|$ has the same parity as an'/k. To see that this is possible, note that $an'/k = an/k - a|V(E^*)|/k$, and $a|V(E_1)|/k - a|V(E_2)|/k = (k-1)a \mod 2$. On the other hand, $|V(E_1) \cap V_1| - |V(E_2) \cap V_1| = (k-1)a + 1 \mod 2$. Since these two differences have different parities, we may choose E^* as required.

Now suppose instead that G has no bad edges. In this case, we take E^* to be empty, and assume that $|V_1'|$ has the same parity as an'/k. In either case we will obtain a contradiction to this choice of E^*, proving that in fact G must have no bad edges and the size of $V_1' = V_1$ has different parity to an/k, and therefore that $\mathcal{P}'$ is the partition we wished to find.

Fix such a matching E^*, and note that $k \mid n'$ since $n' = n - |V(E)|$. We shall find a perfect matching in $G[V_1' \cup V_2']$, that is, covering the vertices which are not covered by E^*. Let $I = \{0 \le i \le k : i = a \mod 2\}$. We next choose numbers $n_i \ge cn$ for $i \in I$, such that $\sum_{i \in I} n_i = n'/k$ and $\sum_{i \in I} in_i = |V_1'|$. To see that this is possible we use a variational argument: we start with all n_i equal to cn except for n_0 or n_1 (according as $0 \in I$ or $1 \in I$), which is chosen so that $\sum_i n_i = n'/k$, so $\sum_{i \in I} in_i$ initially is at most $n'/k + k^2 cn \le |V_1'|$, and has the same parity as $|V_1'|$. We repeatedly decrease some $n_i \ge cn+1$ by 1 and increase n_{i+2} by 1; this increases $\sum_{i \in I} in_i$ by 2, so we eventually achieve $\sum_i in_i = |V_1'|$, as required. Now we choose partitions $\mathcal{P}_1$ of V_1' and $\mathcal{P}_2$ of V_2' uniformly at random from those such that $\mathcal{P}_1$ has n_i parts of size i and $\mathcal{P}_2$ has n_i parts of size $k - i$ for each i. For each $i \in I$, let $X_1^i, \ldots, X_{n_i}^i$ be the parts in $\mathcal{P}_1$ of size i, and $Z_1^i, \ldots, Z_{n_i}^i$ be the parts in $\mathcal{P}_2$ of size $k - i$. Then we may form a partition of $V(G)$ into parts Y_j^i for $i \in I$ and $j \in [k]$ uniformly at random, where for each i and j the part Y_j^i contains one vertex from each X_α^i if $j \le i$, and one vertex from each Z_α^i if $i + 1 \le j$. So each Y_j^i has size $n_i \ge cn$. We consider for each $i \in I$ an auxiliary k-partite k-graph H^i on vertex classes $Y_1^i, \ldots, Y_k^i$ whose edges are those k-partite k-tuples that are edges of G. So the H^i are vertex-disjoint, and to find a perfect matching in G, it suffices to show that with high probability each H^i has a perfect matching.

Fix some $i \in I$. Since we covered all bad vertices by E^*, all vertices in H^i belong to at most $k\sqrt{b}n^{k-1}$ bad edges of G. Now fix $y_1 \in Y_1^i$. Say that a $(k-2)$-tuple $(y_j)_{j=2}^{k-1}$ with $y_j \in Y_j^i$ for $j \in \{2, \ldots, k-1\}$ is y_1-bad if $y_1 \ldots y_{k-1}$ belongs to more than $kb^{1/4}n$ bad edges. Then there are at most $b^{1/4}n^{k-2}$ y_1-bad $(k-2)$-tuples. Suppose that $i < k$, so $Y_k^i \subseteq V_2$; then a k-tuple $y_1 \ldots y_{k-1}x$ is good if $x \in V_2$ or bad if $x \in V_1$. Since $\delta(G) \ge (1/2 - c)n$ and $|V_2| \le (1/2 + c + b)n$, if $(y_i)_{i=2}^{k-1}$ is y_1-good then $y_1 \ldots y_{k-1}x$ is an edge for all but at most $3cn$ vertices $x \in V_2$. The number of these vertices x that lie in Y_k^i is hypergeometric with mean at least $(1 - 3c)|Y_k^i|$, where $|Y_k^i| = n_i \ge cn$. So by the Chernoff bound (Lemma 6.13), with probability $1 - o(1)$ there are at least $(1 - 4c)n_i$ such vertices x for any choice of $y_1 \in Y_1^i$ and y_1-good $(y_i)_{i=2}^{k-1}$. Thus y_1 is contained in at least

$$(n_i^{k-2} - b^{1/4}n^{k-2})(1 - 4c)n_i > (1 - 5c)n_i^{k-1}$$

edges of H^i. By symmetry, with positive probability the same bound holds for all vertices in any H^i (for $i = k$ we have $Y_k \subseteq V_1$, in which case we proceed similarly with the roles of V_1 and V_2 switched). Then each H^i has a perfect matching by Theorem 8.20, contradicting our assumption that G does not have a perfect matching. So we must have the case that no edges of G are bad and V_1 has different parity to an/k, as required. □

CHAPTER 9

The general theory

In this final chapter, we give a more general result, which epitomises the geometric nature of the theory, in that it almost entirely dispenses with degree assumptions. We cannot hope to avoid such assumptions entirely when using methods based on hypergraph regularity, as sparse hypergraphs may have empty reduced systems. Our degree assumptions are as weak as possible within this context, in that the proportional degrees can be $o(1)$ as $n \to \infty$. The hypotheses of our result are framed in terms of the reduced system provided by hypergraph regularity, so it takes a while to set up the statement, and it is not as clean as that of our main theorems (which is why we have deferred the statement until now). However, the extra generality provided by the geometric context will be important for future applications, even in the context of minimum degree thresholds for hypergraph packing problems, where the tight minimum degree may not imply the minimum degree sequence required by our main theorems. (We intend to return to this point in a future paper.) In an attempt to avoid too much generality, we will restrict attention to the non-partite setting here.

First we describe the setting for our theorem. We start with the hypergraph regularity decomposition.

Regularity setting. Let J be a k-complex on n vertices, where $k \mid n$. Let $\mathcal{Q}$ be a balanced partition of $V(J)$ into h parts, and J' be the k-complex of $\mathcal{Q}$-partite edges in J. Let P be an a-bounded ε-regular vertex-equitable $\mathcal{Q}$-partition $(k-1)$-complex on $V(J)$, with clusters $V_1, \ldots, V_{m_1}$ of size n_1. Let G be a $\mathcal{Q}$-partite k-graph on $V(J)$ that is ν-close to J'_k and perfectly ε-regular with respect to P. Let $Z = G \triangle J'_k$.

Next we describe the setting for the reduced system, which names the constructions that were used in the proof of Theorem 7.11. While we do not specify the source of the subsystem (R_0, M_0), it is perhaps helpful to think of it as being randomly chosen so as to inherit the properties of (R, M), as in the proof of Lemma 5.5.

Reduced system setting. Let $R^1 := R_{P\mathcal{Q}}^{J'Z}(\nu, \mathbf{c})$ on $[m_1]$. Also let (R, M) be a matched k-system on $[m]$, where R is the restriction of R^1 to $[m]$. Given any R' and M' which can be formed from R and M respectively by the deletion of the vertices of at most ψm edges of M, and any vertices $u, v \in V(R')$, we will choose $M_0 \subseteq M'$ such that $V_0 := V(M_0)$ satisfies $m_0 := |V_0| \leq m_0^*$ and M_0 includes the edges of M' containing u, v. Let $R_0 = R[V_0]$, $X = X(R_0, M_0) = \{\chi(e) - \chi(e') : e \in R_0, e' \in M_0\} \subseteq \mathbb{R}^{m_0}$ and $\Pi = \{\mathbf{x} \in \mathbb{Z}^{m_0} : \mathbf{x} \cdot \mathbf{1} = 0\}$.

Now we can state our general theorem (in the above setting).

THEOREM 9.1. *Suppose that $k \geq 3$ and $1/n \ll \varepsilon \ll 1/a \ll \nu, 1/h \ll c_k \ll \cdots \ll c_1 \ll \psi \ll \delta, 1/m_0^*, 1/\ell \ll \alpha, \gamma, 1/k$. Suppose that $m \geq (1 - \psi)m_1$, and the following conditions hold.*

(i) *For any R', M', u and v given as above, we may choose M_0 so that (R_0, M_0) satisfies $B(\mathbf{0}, \delta) \cap \Pi \subseteq CH(X)$.*
(ii) $\delta^+(D_\ell(R, M)) \geq \gamma m$.
(iii) *$L_\mathcal{P}(R_k)$ is complete for any partition $\mathcal{P}$ of $V(R)$ into parts of size at least $(\gamma - \alpha)m$.*
(iv) *Every vertex is contained in at least γn^{k-1} edges of J.*

Then J_k contains a perfect matching.

PROOF. We follow the proof of Theorem 7.11, outlining the modifications. Introduce constants with

$$1/n \ll \varepsilon \ll d^* \ll d_a \ll 1/a \ll \nu, 1/h \ll d, c \ll c_k \ll \cdots \ll c_1$$
$$\ll \psi \ll 1/C' \ll 1/B, 1/C \ll \delta, 1/m_0^*, 1/\ell \ll \alpha, \gamma, 1/k.$$

Our set-up already provides Q, J', P, G, Z, R, M, as in that proof. (Note that we are working in the non-partite setting, so there is no $\mathcal{P}$, and F consists of the unique function $f : [k] \to [1]$. We also do not need to consider two reduced systems.) There is no need for an analogue of Claim 7.12 as we already have M. The proof of Claim 7.13 is the same, except that for (iii), instead of the minimum F-degree assumption, we apply Lemma 7.14 to the edges of J containing v, using the assumption that every vertex is contained in at least γn^{k-1} edges of J. To prove Claim 7.15, instead of Lemma 5.7, we first apply Lemma 4.1 to see that (R_0, M_0) is (B, C)-irreducible for any u and v. It follows that (R', M') is (B, C)-irreducible, then we apply Lemma 4.10 to see that $D_{C'}(R', M')$ is complete. The remainder of the proof only uses these claims, so goes through as before. □

Each of the conditions of Theorem 9.1 is necessary for the proof strategy used in Theorem 7.11, so we cannot strengthen this result further using this argument.

We conclude with a remark comparing our techniques to the absorbing method, which has been successfully applied to many hypergraph matching problems. The idea of this method is to randomly select a small matching M_0 in a k-graph G, and show that with high probability it has the property that it can 'absorb' any small set of vertices V_0, in that there is a matching covering $V(M_0) \cup V_0$. Given such a matching M_0, to find a perfect matching in G, it suffices to find an almost-perfect matching in $G \setminus V(M_0)$, which is often a much simpler problem.

For example, the essence of Lo and Markström's independent proof of Theorem 1.2 was to show that, in any k-partite graph G which satisfies the degree conditions of this theorem, a randomly-chosen collection M_0 of εn disjoint copies of K_k in G is absorbing. In this setting this means that for any set $V_0 \subseteq V(G)$ whose size is divisible by k, there exists a perfect K_k-packing in $G[V_0 \cup V(M_0)]$. They then showed that $G \setminus V(M_0)$ must contain an K_k-packing M_1 covering almost all the vertices; the absorbing property of M_1 then implies that $G \setminus V(M_1)$ contains a perfect K_k-packing, which combined with M_1 yields a perfect K_k-packing in G. See [36] for further details.

The advantage of the absorbing method is that it avoids the use of the hypergraph blow-up lemma (and often avoids hypergraph regularity altogether), and so leads to shorter proofs, when it works. However, the existence of an absorbing matching seems to be a fortuitous circumstance in each application of the method, and it is not clear how one could hope to find it in general problems. By contrast, our theory explains 'why' there is a perfect matching, by analysing the only possible obstructions (space barriers and divisibility barriers). This has the additional advantage of giving structural information, so we can apply the stability method to obtain exact results.

Acknowledgements

We would like to thank Allan Lo for pointing out an inconsistency in the proof of Theorem 1.2 in an earlier version, and also an anonymous referee for offering helpful suggestions which have led to an improvement in the presentation of this paper.

Bibliography

[1] Noga Alon, Peter Frankl, Hao Huang, Vojtech Rödl, Andrzej Ruciński, and Benny Sudakov, *Large matchings in uniform hypergraphs and the conjecture of Erdős and Samuels*, J. Combin. Theory Ser. A **119** (2012), no. 6, 1200–1215, DOI 10.1016/j.jcta.2012.02.004. MR2915641

[2] Arash Asadpour, Uriel Feige, and Amin Saberi, *Santa Claus meets hypergraph matchings*, Approximation, randomization and combinatorial optimization, Lecture Notes in Comput. Sci., vol. 5171, Springer, Berlin, 2008, pp. 10–20, DOI 10.1007/978-3-540-85363-3_2. MR2538773

[3] Béla Bollobás, *Combinatorics*, Cambridge University Press, Cambridge, 1986. Set systems, hypergraphs, families of vectors and combinatorial probability. MR866142 (88g:05001)

[4] F. R. K. Chung, R. L. Graham, P. Frankl, and J. B. Shearer, *Some intersection theorems for ordered sets and graphs*, J. Combin. Theory Ser. A **43** (1986), no. 1, 23–37, DOI 10.1016/0097-3165(86)90019-1. MR859293 (87k:05002)

[5] Fan Chung and Linyuan Lu, *An upper bound for the Turán number $t_3(n,4)$*, J. Combin. Theory Ser. A **87** (1999), no. 2, 381–389, DOI 10.1006/jcta.1998.2961. MR1704268 (2000d:05060)

[6] Béla Csaba and Marcelo Mydlarz, *Approximate multipartite version of the Hajnal-Szemerédi theorem*, J. Combin. Theory Ser. B **102** (2012), no. 2, 395–410, DOI 10.1016/j.jctb.2011.10.003. MR2885426 (2012k:05306)

[7] Andrzej Czygrinow and Vikram Kamat, *Tight co-degree condition for perfect matchings in 4-graphs*, Electron. J. Combin. **19** (2012), no. 2, Paper 20, 16. MR2928635

[8] Andrzej Czygrinow and Brendan Nagle, *A note on codegree problems for hypergraphs*, Bull. Inst. Combin. Appl. **32** (2001), 63–69. MR1829685 (2002a:05187)

[9] Andrzej Czygrinow and Brendan Nagle, *On random sampling in uniform hypergraphs*, Random Structures Algorithms **38** (2011), no. 4, 422–440, DOI 10.1002/rsa.20326. MR2829310 (2012f:05265)

[10] David E. Daykin and Roland Häggkvist, *Degrees giving independent edges in a hypergraph*, Bull. Austral. Math. Soc. **23** (1981), no. 1, 103–109, DOI 10.1017/S0004972700006924. MR615135 (82g:05068)

[11] G. A. Dirac, *Some theorems on abstract graphs*, Proc. London Math. Soc. (3) **2** (1952), 69–81. MR0047308 (13,856e)

[12] Jack Edmonds, *Paths, trees, and flowers*, Canad. J. Math. **17** (1965), 449–467. MR0177907 (31 #2165)

[13] Paul Erdős and Miklós Simonovits, *Supersaturated graphs and hypergraphs*, Combinatorica **3** (1983), no. 2, 181–192, DOI 10.1007/BF02579292. MR726456 (85e:05125)

[14] Eldar Fischer, *Variants of the Hajnal-Szemerédi theorem*, J. Graph Theory **31** (1999), no. 4, 275–282, DOI 10.1002/(SICI)1097-0118(199908)31:4⟨275::AID-JGT2⟩3.3.CO;2-6. MR1698745 (2000c:05133)

[15] Peter Frankl and Vojtěch Rödl, *Extremal problems on set systems*, Random Structures Algorithms **20** (2002), no. 2, 131–164, DOI 10.1002/rsa.10017.abs. MR1884430 (2002m:05192)

[16] A. W. Goodman, *On sets of acquaintances and strangers at any party*, Amer. Math. Monthly **66** (1959), 778–783. MR0107610 (21 #6335)

[17] W. T. Gowers, *Hypergraph regularity and the multidimensional Szemerédi theorem*, Ann. of Math. (2) **166** (2007), no. 3, 897–946, DOI 10.4007/annals.2007.166.897. MR2373376 (2000d:05250)

[18] A. Hajnal and E. Szemerédi, *Proof of a conjecture of P. Erdős*, Combinatorial theory and its applications, II (Proc. Colloq., Balatonfüred, 1969), North-Holland, Amsterdam, 1970, pp. 601–623. MR0297607 (45 #6661)

[19] Hiệp Hàn, Yury Person, and Mathias Schacht, *On perfect matchings in uniform hypergraphs with large minimum vertex degree*, SIAM J. Discrete Math. **23** (2009), no. 2, 732–748, DOI 10.1137/080729657. MR2496914 (2011a:05266)
[20] Svante Janson, Tomasz Łuczak, and Andrzej Ruciński, *Random graphs*, Wiley-Interscience Series in Discrete Mathematics and Optimization, Wiley-Interscience, New York, 2000. MR1782847 (2001k:05180)
[21] Richard M. Karp, *Reducibility among combinatorial problems*, Complexity of computer computations (Proc. Sympos., IBM Thomas J. Watson Res. Center, Yorktown Heights, N.Y., 1972), Plenum, New York, 1972, pp. 85–103. MR0378476 (51 #14644)
[22] Peter Keevash, *A hypergraph blow-up lemma*, Random Structures Algorithms **39** (2011), no. 3, 275–376, DOI 10.1002/rsa.20362. MR2816936
[23] Peter Keevash, *Hypergraph Turán problems*, Surveys in combinatorics 2011, London Math. Soc. Lecture Note Ser., vol. 392, Cambridge Univ. Press, Cambridge, 2011, pp. 83–139. MR2866732
[24] P. Keevash, F. Knox and R. Mycroft, Polynomial-time perfect matchings in dense hypergraphs, *Proc. 45th STOC*, to appear.
[25] Peter Keevash, Daniela Kühn, Richard Mycroft, and Deryk Osthus, *Loose Hamilton cycles in hypergraphs*, Discrete Math. **311** (2011), no. 7, 544–559, DOI 10.1016/j.disc.2010.11.013. MR2765622 (2012d:05274)
[26] P. Keevash and R. Mycroft, A multipartite Hajnal-Szemerédi theorem, arXiv:1201.1882.
[27] I. Khan, Perfect Matching in 3 uniform hypergraphs with large vertex degree, arXiv:1101.5830.
[28] I. Khan, Perfect matchings in 4-uniform hypergraphs, arXiv:1101.5675.
[29] János Komlós, Gábor N. Sárközy, and Endre Szemerédi, *Blow-up lemma*, Combinatorica **17** (1997), no. 1, 109–123, DOI 10.1007/BF01196135. MR1466579 (99b:05083)
[30] Daniela Kühn and Deryk Osthus, *Matchings in hypergraphs of large minimum degree*, J. Graph Theory **51** (2006), no. 4, 269–280, DOI 10.1002/jgt.20139. MR2207573 (2006k:05177)
[31] Daniela Kühn and Deryk Osthus, *The minimum degree threshold for perfect graph packings*, Combinatorica **29** (2009), no. 1, 65–107, DOI 10.1007/s00493-009-2254-3. MR2506388 (2010i:05271)
[32] Daniela Kühn and Deryk Osthus, *Embedding large subgraphs into dense graphs*, Surveys in combinatorics 2009, London Math. Soc. Lecture Note Ser., vol. 365, Cambridge Univ. Press, Cambridge, 2009, pp. 137–167. MR2588541 (2011c:05275)
[33] Daniela Kühn, Deryk Osthus, and Andrew Treglown, *Matchings in 3-uniform hypergraphs*, J. Combin. Theory Ser. B **103** (2013), no. 2, 291–305, DOI 10.1016/j.jctb.2012.11.005. MR3018071
[34] Marek Karpiński, Andrzej Ruciński, and Edyta Szymańska, *The complexity of perfect matching problems on dense hypergraphs*, Algorithms and computation, Lecture Notes in Comput. Sci., vol. 5878, Springer, Berlin, 2009, pp. 626–636, DOI 10.1007/978-3-642-10631-6_64. MR2792760 (2012c:05244)
[35] A. Lo and K. Markström, F-factors in hypergraphs via absorption, arXiv:1105.3411.
[36] A. Lo and K. Markström, A multipartite version of the Hajnal-Szemerédi theorem for graphs and hypergraphs, arXiv:1108.4184.
[37] L. Lovász and M. Simonovits, *On the number of complete subgraphs of a graph. II*, Studies in pure mathematics, Birkhäuser, Basel, 1983, pp. 459–495. MR820244 (87a:05089)
[38] Csaba Magyar and Ryan R. Martin, *Tripartite version of the Corrádi-Hajnal theorem*, Discrete Math. **254** (2002), no. 1-3, 289–308, DOI 10.1016/S0012-365X(01)00373-9. MR1910115 (2003f:05073)
[39] Klas Markström and Andrzej Ruciński, *Perfect matchings (and Hamilton cycles) in hypergraphs with large degrees*, European J. Combin. **32** (2011), no. 5, 677–687, DOI 10.1016/j.ejc.2011.02.001. MR2788783 (2012c:05247)
[40] Ryan Martin and Endre Szemerédi, *Quadripartite version of the Hajnal-Szemerédi theorem*, Discrete Math. **308** (2008), no. 19, 4337–4360, DOI 10.1016/j.disc.2007.08.019. MR2433861 (2009d:05201)
[41] Dhruv Mubayi and Vojtěch Rödl, *Uniform edge distribution in hypergraphs is hereditary*, Electron. J. Combin. **11** (2004), no. 1, Research Paper 55, 32. MR2097321 (2005i:05092)
[42] R. Mycroft, Packing k-partite k-graphs, in preparation.

[43] Brendan Nagle and Vojtěch Rödl, *Regularity properties for triple systems*, Random Structures Algorithms **23** (2003), no. 3, 264–332, DOI 10.1002/rsa.10094. MR1999038 (2004e:05098)

[44] Oleg Pikhurko, *Perfect matchings and K_4^3-tilings in hypergraphs of large codegree*, Graphs Combin. **24** (2008), no. 4, 391–404, DOI 10.1007/s00373-008-0787-7. MR2438870 (2009e:05214)

[45] Alexander A. Razborov, *On the minimal density of triangles in graphs*, Combin. Probab. Comput. **17** (2008), no. 4, 603–618, DOI 10.1017/S0963548308009085. MR2433944 (2009i:05118)

[46] Vojtech Rödl and Andrzej Ruciński, *Dirac-type questions for hypergraphs—a survey (or more problems for Endre to solve)*, An irregular mind, Bolyai Soc. Math. Stud., vol. 21, János Bolyai Math. Soc., Budapest, 2010, pp. 561–590, DOI 10.1007/978-3-642-14444-8_16. MR2815614 (2012j:05008)

[47] Vojtěch Rödl, Andrzej Ruciński, and Endre Szemerédi, *A Dirac-type theorem for 3-uniform hypergraphs*, Combin. Probab. Comput. **15** (2006), no. 1-2, 229–251, DOI 10.1017/S0963548305007042. MR2195584 (2006j:05144)

[48] Vojtech Rödl, Andrzej Ruciński, and Endre Szemerédi, *Perfect matchings in uniform hypergraphs with large minimum degree*, European J. Combin. **27** (2006), no. 8, 1333–1349, DOI 10.1016/j.ejc.2006.05.008. MR2260124 (2007g:05153)

[49] Vojtech Rödl, Andrzej Ruciński, and Endre Szemerédi, *Perfect matchings in large uniform hypergraphs with large minimum collective degree*, J. Combin. Theory Ser. A **116** (2009), no. 3, 613–636, DOI 10.1016/j.jcta.2008.10.002. MR2500161 (2010d:05124)

[50] Vojtěch Rödl and Mathias Schacht, *Regular partitions of hypergraphs: regularity lemmas*, Combin. Probab. Comput. **16** (2007), no. 6, 833–885. MR2351688 (2008h:05083)

[51] Vojtěch Rödl and Mathias Schacht, *Property testing in hypergraphs and the removal lemma [extended abstract]*, STOC'07—Proceedings of the 39th Annual ACM Symposium on Theory of Computing, ACM, New York, 2007, pp. 488–495, DOI 10.1145/1250790.1250862. MR2402474 (2009f:68204)

[52] Vojtěch Rödl and Jozef Skokan, *Regularity lemma for k-uniform hypergraphs*, Random Structures Algorithms **25** (2004), no. 1, 1–42, DOI 10.1002/rsa.20017. MR2069663 (2005d:05144)

[53] Alexander Schrijver, *Theory of linear and integer programming*, Wiley-Interscience Series in Discrete Mathematics, John Wiley & Sons Ltd., Chichester, 1986. A Wiley-Interscience Publication. MR874114 (88m:90090)

[54] Benny Sudakov and Jan Vondrák, *A randomized embedding algorithm for trees*, Combinatorica **30** (2010), no. 4, 445–470, DOI 10.1007/s00493-010-2422-5. MR2728498 (2012c:05340)

[55] Endre Szemerédi, *Regular partitions of graphs* (English, with French summary), Problèmes combinatoires et théorie des graphes (Colloq. Internat. CNRS, Univ. Orsay, Orsay, 1976), Colloq. Internat. CNRS, vol. 260, CNRS, Paris, 1978, pp. 399–401. MR540024 (81i:05095)

[56] Edyta Szymańska, *The complexity of almost perfect matchings in uniform hypergraphs with high codegree*, Combinatorial algorithms, Lecture Notes in Comput. Sci., vol. 5874, Springer, Berlin, 2009, pp. 438–449, DOI 10.1007/978-3-642-10217-2_43. MR2577959

[57] Andrew Treglown and Yi Zhao, *Exact minimum degree thresholds for perfect matchings in uniform hypergraphs*, J. Combin. Theory Ser. A **119** (2012), no. 7, 1500–1522, DOI 10.1016/j.jcta.2012.04.006. MR2925939

[58] P. Turán, *On an extremal problem in graph theory* (in Hungarian), Mat. Fiz. Lapok **48** (1941), 436–452.

Editorial Information

To be published in the *Memoirs*, a paper must be correct, new, nontrivial, and significant. Further, it must be well written and of interest to a substantial number of mathematicians. Piecemeal results, such as an inconclusive step toward an unproved major theorem or a minor variation on a known result, are in general not acceptable for publication.

Papers appearing in *Memoirs* are generally at least 80 and not more than 200 published pages in length. Papers less than 80 or more than 200 published pages require the approval of the Managing Editor of the Transactions/Memoirs Editorial Board. Published pages are the same size as those generated in the style files provided for $\mathcal{AMS}$-LATEX or $\mathcal{AMS}$-TEX.

Information on the backlog for this journal can be found on the AMS website starting from http://www.ams.org/memo.

A Consent to Publish is required before we can begin processing your paper. After a paper is accepted for publication, the Providence office will send a Consent to Publish and Copyright Agreement to all authors of the paper. By submitting a paper to the *Memoirs*, authors certify that the results have not been submitted to nor are they under consideration for publication by another journal, conference proceedings, or similar publication.

Information for Authors

Memoirs is an author-prepared publication. Once formatted for print and on-line publication, articles will be published as is with the addition of AMS-prepared frontmatter and backmatter. Articles are not copyedited; however, confirmation copy will be sent to the authors.

Initial submission. The AMS uses Centralized Manuscript Processing for initial submissions. Authors should submit a PDF file using the Initial Manuscript Submission form found at www.ams.org/submission/memo, or send one copy of the manuscript to the following address: Centralized Manuscript Processing, MEMOIRS OF THE AMS, 201 Charles Street, Providence, RI 02904-2294 USA. If a paper copy is being forwarded to the AMS, indicate that it is for *Memoirs* and include the name of the corresponding author, contact information such as email address or mailing address, and the name of an appropriate Editor to review the paper (see the list of Editors below).

The paper must contain a *descriptive title* and an *abstract* that summarizes the article in language suitable for workers in the general field (algebra, analysis, etc.). The *descriptive title* should be short, but informative; useless or vague phrases such as "some remarks about" or "concerning" should be avoided. The *abstract* should be at least one complete sentence, and at most 300 words. Included with the footnotes to the paper should be the 2010 *Mathematics Subject Classification* representing the primary and secondary subjects of the article. The classifications are accessible from www.ams.org/msc/. The Mathematics Subject Classification footnote may be followed by a list of *key words and phrases* describing the subject matter of the article and taken from it. Journal abbreviations used in bibliographies are listed in the latest *Mathematical Reviews* annual index. The series abbreviations are also accessible from www.ams.org/msnhtml/serials.pdf. To help in preparing and verifying references, the AMS offers MR Lookup, a Reference Tool for Linking, at www.ams.org/mrlookup/.

Electronically prepared manuscripts. The AMS encourages electronically prepared manuscripts, with a strong preference for $\mathcal{AMS}$-LATEX. To this end, the Society has prepared $\mathcal{AMS}$-LATEX author packages for each AMS publication. Author packages include instructions for preparing electronic manuscripts, samples, and a style file that generates the particular design specifications of that publication series. Though $\mathcal{AMS}$-LATEX is the highly preferred format of TEX, author packages are also available in $\mathcal{AMS}$-TEX.

Authors may retrieve an author package for *Memoirs of the AMS* from www.ams.org/journals/memo/memoauthorpac.html or via FTP to ftp.ams.org (login as anonymous, enter your complete email address as password, and type cd pub/author-info). The

AMS Author Handbook and the *Instruction Manual* are available in PDF format from the author package link. The author package can also be obtained free of charge by sending email to `tech-support@ams.org` or from the Publication Division, American Mathematical Society, 201 Charles St., Providence, RI 02904-2294, USA. When requesting an author package, please specify $\mathcal{AMS}$-L^{A}T$_E$X or $\mathcal{AMS}$-T$_E$X and the publication in which your paper will appear. Please be sure to include your complete mailing address.

After acceptance. The source files for the final version of the electronic manuscript should be sent to the Providence office immediately after the paper has been accepted for publication. The author should also submit a PDF of the final version of the paper to the editor, who will forward a copy to the Providence office.

Accepted electronically prepared files can be submitted via the web at `www.ams.org/submit-book-journal/`, sent via FTP, or sent on CD to the Electronic Prepress Department, American Mathematical Society, 201 Charles Street, Providence, RI 02904-2294 USA. T$_E$X source files and graphic files can be transferred over the Internet by FTP to the Internet node `ftp.ams.org` (130.44.1.100). When sending a manuscript electronically via CD, please be sure to include a message indicating that the paper is for the *Memoirs*.

Electronic graphics. Comprehensive instructions on preparing graphics are available at `www.ams.org/authors/journals.html`. A few of the major requirements are given here.

Submit files for graphics as EPS (Encapsulated PostScript) files. This includes graphics originated via a graphics application as well as scanned photographs or other computer-generated images. If this is not possible, TIFF files are acceptable as long as they can be opened in Adobe Photoshop or Illustrator.

Authors using graphics packages for the creation of electronic art should also avoid the use of any lines thinner than 0.5 points in width. Many graphics packages allow the user to specify a "hairline" for a very thin line. Hairlines often look acceptable when proofed on a typical laser printer. However, when produced on a high-resolution laser imagesetter, hairlines become nearly invisible and will be lost entirely in the final printing process.

Screens should be set to values between 15% and 85%. Screens which fall outside of this range are too light or too dark to print correctly. Variations of screens within a graphic should be no less than 10%.

Inquiries. Any inquiries concerning a paper that has been accepted for publication should be sent to `memo-query@ams.org` or directly to the Electronic Prepress Department, American Mathematical Society, 201 Charles St., Providence, RI 02904-2294 USA.

Editors

This journal is designed particularly for long research papers, normally at least 80 pages in length, and groups of cognate papers in pure and applied mathematics. Papers intended for publication in the *Memoirs* should be addressed to one of the following editors. The AMS uses Centralized Manuscript Processing for initial submissions to AMS journals. Authors should follow instructions listed on the Initial Submission page found at www.ams.org/memo/memosubmit.html.

Algebra, to ALEXANDER KLESHCHEV, Department of Mathematics, University of Oregon, Eugene, OR 97403-1222; e-mail: klesh@uoregon.edu

Algebraic geometry, to DAN ABRAMOVICH, Department of Mathematics, Brown University, Box 1917, Providence, RI 02912; e-mail: amsedit@math.brown.edu

Algebraic topology, to SOREN GALATIUS, Department of Mathematics, Stanford University, Stanford, CA 94305 USA; e-mail: transactions@lists.stanford.edu

Arithmetic geometry, to TED CHINBURG, Department of Mathematics, University of Pennsylvania, Philadelphia, PA 19104-6395; e-mail: math-tams@math.upenn.edu

Automorphic forms, representation theory and combinatorics, to DANIEL BUMP, Department of Mathematics, Stanford University, Building 380, Sloan Hall, Stanford, California 94305; e-mail: bump@math.stanford.edu

Combinatorics and discrete geometry, to IGOR PAK, Department of Mathematics, University of California, Los Angeles, California 90095; e-mail: pak@math.ucla.edu

Commutative and homological algebra, to LUCHEZAR L. AVRAMOV, Department of Mathematics, University of Nebraska, Lincoln, NE 68588-0130; e-mail: avramov@math.unl.edu

Differential geometry and global analysis, to CHRIS WOODWARD, Department of Mathematics, Rutgers University, 110 Frelinghuysen Road, Piscataway, NJ 08854; e-mail: ctw@math.rutgers.edu

Dynamical systems and ergodic theory and complex analysis, to YUNPING JIANG, Department of Mathematics, CUNY Queens College and Graduate Center, 65-30 Kissena Blvd., Flushing, NY 11367; e-mail: Yunping.Jiang@qc.cuny.edu

Ergodic theory and combinatorics, to VITALY BERGELSON, Ohio State University, Department of Mathematics, 231 W. 18th Ave, Columbus, OH 43210; e-mail: vitaly@math.ohio-state.edu

Functional analysis and operator algebras, to NATHANIEL BROWN, Department of Mathematics, 320 McAllister Building, Penn State University, University Park, PA 16802; e-mail: nbrown@math.psu.edu

Geometric analysis, to WILLIAM P. MINICOZZI II, Department of Mathematics, Johns Hopkins University, 3400 N. Charles St., Baltimore, MD 21218; e-mail: trans@math.jhu.edu

Geometric topology, to MARK FEIGHN, Math Department, Rutgers University, Newark, NJ 07102; e-mail: feighn@andromeda.rutgers.edu

Harmonic analysis, complex analysis, to MALABIKA PRAMANIK, Department of Mathematics, 1984 Mathematics Road, University of British Columbia, Vancouver, BC, Canada V6T 1Z2; e-mail: malabika@math.ubc.ca

Harmonic analysis, representation theory, and Lie theory, to E. P. VAN DEN BAN, Department of Mathematics, Utrecht University, P.O. Box 80 010, 3508 TA Utrecht, The Netherlands; e-mail: E.P.vandenBan@uu.nl

Logic, to ANTONIO MONTALBAN, Department of Mathematics, The University of California, Berkeley, Evans Hall #3840, Berkeley, California, CA 94720; e-mail: antonio@math.berkeley.edu

Number theory, to SHANKAR SEN, Department of Mathematics, 505 Malott Hall, Cornell University, Ithaca, NY 14853; e-mail: ss70@cornell.edu

Partial differential equations, to MARKUS KEEL, School of Mathematics, University of Minnesota, Minneapolis, MN 55455; e-mail: keel@math.umn.edu

Partial differential equations and functional analysis, to ALEXANDER KISELEV, Department of Mathematics, University of Wisconsin-Madison, 480 Lincoln Dr., Madison, WI 53706; e-mail: kisilev@math.wisc.edu

Probability and statistics, to PATRICK FITZSIMMONS, Department of Mathematics, University of California, San Diego, 9500 Gilman Drive, La Jolla, CA 92093-0112; e-mail: pfitzsim@math.ucsd.edu

Real analysis and partial differential equations, to WILHELM SCHLAG, Department of Mathematics, The University of Chicago, 5734 South University Avenue, Chicago, IL 60615; e-mail: schlag@math.uchicago.edu

All other communications to the editors, should be addressed to the Managing Editor, ALEJANDRO ADEM, Department of Mathematics, The University of British Columbia, Room 121, 1984 Mathematics Road, Vancouver, B.C., Canada V6T 1Z2; e-mail: adem@math.ubc.ca

SELECTED PUBLISHED TITLES IN THIS SERIES

1094 **Ian F. Putnam,** A Homology Theory for Smale Spaces, 2014
1093 **Ron Blei,** The Grothendieck Inequality Revisited, 2014
1092 **Yun Long, Asaf Nachmias, Weiyang Ning, and Yuval Peres,** A Power Law of Order 1/4 for Critical Mean Field Swendsen-Wang Dynamics, 2014
1091 **Vilmos Totik,** Polynomial Approximation on Polytopes, 2014
1090 **Ameya Pitale, Abhishek Saha, and Ralf Schmidt,** Transfer of Siegel Cusp Forms of Degree 2, 2014
1089 **Peter Šemrl,** The Optimal Version of Hua's Fundamental Theorem of Geometry of Rectangular Matrices, 2014
1088 **Mark Green, Phillip Griffiths, and Matt Kerr,** Special Values of Automorphic Cohomology Classes, 2014
1087 **Colin J. Bushnell and Guy Henniart,** To an Effective Local Langlands Correspondence, 2014
1086 **Stefan Ivanov, Ivan Minchev, and Dimiter Vassilev,** Quaternionic Contact Einstein Structures and the Quaternionic Contact Yamabe Problem, 2014
1085 **A. L. Carey, V. Gayral, A. Rennie, and F. A. Sukochev,** Index Theory for Locally Compact Noncommutative Geometries, 2014
1084 **Michael S. Weiss and Bruce E. Williams,** Automorphisms of Manifolds and Algebraic K-Theory: Part III, 2014
1083 **Jakob Wachsmuth and Stefan Teufel,** Effective Hamiltonians for Constrained Quantum Systems, 2014
1082 **Fabian Ziltener,** A Quantum Kirwan Map: Bubbling and Fredholm Theory for Symplectic Vortices over the Plane, 2014
1081 **Sy-David Friedman, Tapani Hyttinen, and Vadim Kulikov,** Generalized Descriptive Set Theory and Classification Theory, 2014
1080 **Vin de Silva, Joel W. Robbin, and Dietmar A. Salamon,** Combinatorial Floer Homology, 2014
1079 **Pascal Lambrechts and Ismar Volić,** Formality of the Little N-disks Operad, 2013
1078 **Milen Yakimov,** On the Spectra of Quantum Groups, 2013
1077 **Christopher P. Bendel, Daniel K. Nakano, Brian J. Parshall, and Cornelius Pillen,** Cohomology for Quantum Groups via the Geometry of the Nullcone, 2013
1076 **Jaeyoung Byeon and Kazunaga Tanaka,** Semiclassical Standing Waves with Clustering Peaks for Nonlinear Schrödinger Equations, 2013
1075 **Deguang Han, David R. Larson, Bei Liu, and Rui Liu,** Operator-Valued Measures, Dilations, and the Theory of Frames, 2013
1074 **David Dos Santos Ferreira and Wolfgang Staubach,** Global and Local Regularity of Fourier Integral Operators on Weighted and Unweighted Spaces, 2013
1073 **Hajime Koba,** Nonlinear Stability of Ekman Boundary Layers in Rotating Stratified Fluids, 2014
1072 **Victor Reiner, Franco Saliola, and Volkmar Welker,** Spectra of Symmetrized Shuffling Operators, 2014
1071 **Florin Diacu,** Relative Equilibria in the 3-Dimensional Curved n-Body Problem, 2014
1070 **Alejandro D. de Acosta and Peter Ney,** Large Deviations for Additive Functionals of Markov Chains, 2014
1069 **Ioan Bejenaru and Daniel Tataru,** Near Soliton Evolution for Equivariant Schrödinger Maps in Two Spatial Dimensions, 2014
1068 **Florica C. Cîrstea,** A Complete Classification of the Isolated Singularities for Nonlinear Elliptic Equations with Inverse Square Potentials, 2014

For a complete list of titles in this series, visit the
AMS Bookstore at **www.ams.org/bookstore/memoseries/**.